Eco-Design and Ecological Transition

Smart Innovation Set

coordinated by
Dimitri Uzunidis

Volume 42

Eco-Design and Ecological Transition

Questioning the Economic Model

Marie-France Vernier

iSTE

WILEY

First published 2023 in Great Britain and the United States by ISTE Ltd and John Wiley & Sons, Inc.

ISTE Ltd
27-37 St George's Road
London SW19 4EU
UK

www.iste.co.uk

John Wiley & Sons, Inc.
111 River Street
Hoboken, NJ 07030
USA

www.wiley.com

Library of Congress Control Number: 2023943095

British Library Cataloguing-in-Publication Data
A CIP record for this book is available from the British Library
ISBN 978-1-78630-804-7

Contents

Foreword

An economic model on its last legs, suffocated by a pollution that is difficult to curb. This is the starting point of Marie-France Vernier's reflection in her book *Eco-design and Ecological Transition*. This economic model, capitalism, is based on the realization of a process of accumulation in which the investment, combining material capital and labor force, must give rise to a production of a greater value, a source of profit at the time of its market valuation. This accumulation process deploys a linear economy in which natural resources are considered almost unlimited and the emphasis is on the production of value during the production process and its market realization. In the same way, once the profit is made by the act of selling, the future of the goods leaves the field of interest of the economic actors. In other words, in the "extract–produce–consume–discard" sequence, the two extremities have been little considered for a long time. The theorists of this economic model nevertheless sensed the untenable lightness of such a sequence, emphasizing for some the insufficiency of natural resources to provide for the needs of a growing population (Malthus), or the rupture between human and nature through industrialization (Marx), or the costs induced by too much pollution (Pigou), etc. But these presentiments did not emerge from the carefree enthusiasm provoked by the new technologies and by the improvement of well-being following the diffusion of all kinds of goods that flooded the countries at the heart of the first industrial revolutions.

In this sequence, the "production" and "consumption" steps have guided the strategies of the actors of production and consumption, often with the ambition of increasing their scope and intensity. They have been at the center of thinkers' reflections through, for example, highlighting the sources of the increase in labor productivity, and also of the consequences of exploiting labor power, or of the limits of the relationship between consumption and well-being. But the shortcomings that have been highlighted, especially concerning economic and social inequalities, concretized by the regular social protests, have not been the first instigator of the current collective reflection on the necessity of an adaptation or a change of economic model.

The origin of the current questioning of the economic model results more from the division provoked by the two ends of the sequence, "extract" and "discard", and its consequences on the heart that constitute production and consumption. The awareness of the non-renewable and therefore finite nature of a certain number of resources essential to the production process sounded at first like a faraway threat, until it became an annual countdown moving ever closer to the "day of overshoot" (July 28 in 2022, according to the calculations of the Global Footprint Network), from which humanity would draw irreversibly on the "non-renewable" resources of the Earth. At the same time, the release into the atmosphere of the residues of production based on fossil fuels accentuates a climate change that creates uncertainties. Alongside this, the accumulation of material waste, long hidden from the eyes of the Western world in the "pollution havens" created by globalization, is suddenly resurfacing, for example in a seventh continent, a "plastic vortex", estimated at 1.6 million km^2 and located in the northeast Pacific Ocean.

The neglect of these two sequences, "extracting" and "discarding", creates costs that weigh on the core of the economic model, "produce" and "consume". Paradoxically, they are the driving force behind the change or the necessary "transition" of the economic model. Marie-France Vernier deals with this transition by starting from the heart – production – and analyzes with finesse the positive discussion of modifying the production process and the design of goods and services, a modification that could be a vector of economic and social renewal. How to take into account the environmental question in production? If the first strategies of companies, guided by environmental regulations, were to try to reduce the environmental impacts at the "end of the chain", eco-design, or even taking into account of the environmental impact at each stage of the production process of a good, quickly appeared as a solution for everything to change – without anything actually changing.

Indeed, Marie-France Vernier expresses it very well, eco-design has imposed itself in a minimalist version, "technocentric", in a belief that technology could replace the other forms of destroyed or missing capital, notably natural capital, the fable of the drone replacing the bee in its work of pollination. Despite the availability of rigorous tools, such as life cycle assessments (LCAs), the reduction of the environmental impact is sometimes too restricted to one stage of the production process, or even leads to higher costs at other stages. It is also mainly this type of eco-design that is associated with the circular economy. This can be defined as an economic system, which is based on a set of practices designed to better manage the "extract" and "throw away" sequences by reducing the resources used and introducing reduce, reuse and recycling loops in the production and consumption processes. The potential for change associated with (technocentric) eco-design, despite the technological, organizational and commercial innovations to which it gives rise, thus remains low.

A "more sustainable conception", which would integrate not only environmental but also social dimensions into the analysis, implying a rethinking of the relationship between human, the object and the environment, would be more promising. It would imply large-scale, systemic transformations, which would not only seek to limit the negative externalities of the linear economy, as we have defined it, but above all to rethink the ways of "meeting the needs of current generations without jeopardizing the ability of future generations to meet their own needs".

Marie-France Vernier outlines a few avenues, such as a more collective (open) conception and management of resources in the form of the "commons". As a sort of third way between public and private property, the commons corresponds to the definition of forms of collective use and management of resources by communities of actors with the aim of preventing server and perpetuating these resources. These avenues will certainly be a source of reflection and perhaps of creativity for the readers of Marie-France Vernier's work.

Blandine LAPERCHE
Professor of Economics
Université du Littoral-Côte-d'Opale
Chair of the Innovation Research Network

Introduction

The ability of companies to lead a transition to a more sustainable and environmentally friendly economy is an increasingly important issue for our society. The launch of the Green Pact for Europe by the European Commission in 2019 is symptomatic: the announced objective is a climate-neutral Europe with, in particular, the end of greenhouse gas emissions, a growth based on a "clean and circular" economy and actions to fight against biodiversity loss. In France, the 2015 law on the energy transition for green growth mentions the terms transition and sufficiency.

Although there is a lot of talk about companies' commitments to sustainable development and green or ecological products, production, consumption and pollution are still increasing. In Europe, for example, the total volume of waste increased by 14% between 2004 and 2018 according to Eurostat data (2022)[1], while over the same period the European population increased by 2.9%. Similarly, according to "The Global E-waste Monitor" report, published in partnership with the United Nations Institute for Training and Research (UNITAR), e-waste is expected to increase from 53.6 million tons in 2019 to 74 million tons in 2030, an expected increase of 38%! According to the same report, the collection and recycling rate is only 17.4% to date.

1 See: https://ec.europa.eu/.

The ecological transition is still far off or in its infancy. It requires companies to adapt their strategies. At the heart of our economic model, they are in close contact with several players: consumers, distributors, suppliers, and also local authorities and training centers. They are at the center of the exchange of goods and services on which we depend every day.

Eco-design can be a response by allowing the company to offer products or services with a lower environmental impact. It thus contributes to the ecological transition. This book proposes an analysis and a perspective implementation from a reflection at the heart of the products and services of our everyday lives. We begin by defining the central terms of the debate: economic model, ecological transition and eco-design.

Elements of definition

The current economic model

We use the term economic model to designate the dominant mode of organization in our societies in Europe and in a large part of the American and Asian continents. It has several characteristics. It is based on market relations in the areas of production and consumption. Individuals buy or sell goods or services for food, leisure or health. This consumption is based on a system of production in which capital, labor and natural resources are transformed into products on the basis of a capitalist system. The private ownership of the means of production gives great power to the owners of capital, who decide on the major orientations of companies. The search for, or at least the maintenance of, profitability is a determining element of these strategic choices (Beaud 2010; Bardelli 2016).

With the first industrial revolution, this mode of production allowed for decisive progress in the areas of health, food and housing, as well as for increasingly visible pollutions in the environmental field (Beaud 2010). This mode of production irrevocably pollutes and damages the environment. It consumes large quantities of renewable and non-renewable natural resources. The former is endangered by pollution and the latter is lost once used.

This economic model is also characterized by a globalization of the exchange of goods, capital and services. It means an "interdependence of the elements (societies, nations, firms, cultures) that constitute a complex system" (Bocquet 2018, p. 18). Companies are thus dependent on their suppliers for raw materials and semi-finished products. At the same time, they generate pollution that is dispersed to the four corners of the planet. As early as the 1970s, the environmental alarm was sounded with the publication of the Meadows report at the Stockholm conference. As Olivier Godart (2015, p. 64) writes:

> There is the discovery of the feedback effects of man's ability to alter, transform, develop his biophysical environment well beyond the model of adaptation to a given environment.

The social consequences of the current economic model should not be forgotten (from the putting to work of the most fragile in the factories and the development of wage-earning with the first industrial revolution to the new forms of flexibility of work today); nevertheless, this work is centered on the environmental stakes.

The necessary ecological transition

The term ecological transition reveals the growing dissatisfaction of our society with the disastrous environmental and social consequences for the planet, its fauna and flora and also for society itself. It is a "dual awareness of the finiteness of the resources on which global economic growth is based and of the environmental footprint resulting from our energy consumption" (Monnoyer-Smith 2017, p. 5).

Transition involves economic and social changes. It thus designates a process leading to a new model. It requires a mutation, according to De Perthuis (2014), in order to reconcile the current economic model with “the great cycles of natural regulation that allow the reproduction of resources”. According to ADEME, “the ecological transition implies major changes in production systems and lifestyles, as well as the consideration of economic and social impacts, and more broadly, multiple conditions of feasibility” (Vidalenc et al. 2022, p. 9).

The transition is not only ecological, but also social, as it means a change from an unsustainable production and consumption model to a socio-economic model with an impact on the environment that is “acceptable” or sustainable in the long run (Stamm 2015). The environment is then considered as a common good to be preserved for the well-being of all. Transition then consists of renewing ways of consuming, producing, working and living together (Serveille et al. 2017). Some authors are more radical: it must lead “to a more sober system, and, as a result, likely to be more sustainable, equitable and stable” (Bourg and Papaux 2015, p. 1007).

The transition would be unavoidable as soon as society no longer has the resources to live as it has for several centuries. The transition can be undergone or chosen. If it is undergone, society is obliged to adapt to the scarcity of resources, climate change, the loss of biodiversity, etc. These changes create tensions, as individuals are unequal in their access to natural resources such as water or soil, or in the consequences of natural disasters (Bourg and Papaux 2015). Chosen, society forges its destiny: with major changes in modes of production and consumption, and thus challenging established systemic relationships, with the ultimate goal of enabling future generations to meet their needs. States, through public policies, are beginning to guide actors in this direction. Some consumers are opting for products with less environmental impact, or for bulk purchases in order to drastically reduce their waste. Companies, sometimes under public pressure and sometimes voluntarily with strategies of corporate social responsibility (CSR), are modifying their production to limit their environmental impact. In other words, individuals, in their consumption decisions or within a company, adopt an ethical approach and support the ecological transition. It is ethical because it reflects “the search for a good life, with and for others, in just institutions” (Ricoeur 1990, p. 202).

Eco-design: an answer?

Eco-design enables innovation by integrating the environmental impact as early as possible in the design process. It takes place during the design phase, from the emergence of the idea to the detailed definition of the final product or service. This pre-production phase can be a source of innovation and contribute to the ecological transition.

In the economic literature, innovation is most often seen as a source of economic progress and growth. It is also at the heart of the question of our society's capacity to change its economic model, because individuals can introduce changes with environmental and social consequences. Innovation consists of introducing a new or significantly improved product, marketing or organizational mode to the market. Since the work of Joseph Schumpeter in 1934, innovation has traditionally been considered the central factor in the development of business competitiveness and economic growth.

In the early 1960s, authors focused on environmental innovations. These are socio-technical solutions that preserve resources, reduce environmental degradation and/or recover the value of substances already used in economic activity (De Jesus and Mendonça 2018). Habitually, the literature on eco-innovation distinguishes three main types of motivations: they can be regulatory, voluntary or market-driven (Horbach et al. 2012; Rexhäuser and Rammer 2014).

In addition to the technological aspects, which are very present in the literature, environmental innovations can have organizational, social or institutional dimensions (Rennings 2000; OECD 2009). They are based, for example, on new uses, such as the functional economy.

In the literature, eco-design is associated with the circular economy (Vence and Pereira 2019; Garrido and Prada 2021) as a step before the circular economy to define environmental issues or as a component of the circular economy itself (Ghisetti and Montresor 2020). The functional economy and frugal innovations also contribute to the reduction of ecological impact.

Purpose and structure of the book

The purpose of this book is to understand how eco-design can propose solutions to the environmental limitations of our economic model and thus contribute to the ecological transition. We ask ourselves in particular how the environmental and social changes allowed by the eco-design method are or are not sufficient to modify the current economic system.

This book is part of an evolutionary and systemic perspective. Eco-design is a method that appeared in the 1960s, which has evolved considerably in a historical context in which environmental issues have become increasingly important, and even restrictive, for companies under pressure from public policies and their customers. These multiple demands lead to irreversible changes (Nelson and Winter 1982): eco-design is thus a term increasingly used by companies to demonstrate their environmental approach. It thus fuels a process of environmental and social innovation that can contribute to the ecological transition.

From a systemic perspective, we will study the conditions explaining the diffusion of eco-design and its consequences in firms. We are thus interested in the system as an environment in which firms evolve and which impacts their strategies (Coppin 2002). Our analysis focuses on the actors of the socio-technical system, the relationships they maintain and the rules that determine the dynamics of this system (Geels 2004; Hekkert et al. 2007). We can then understand the consequences for companies in terms of practices and economic results.

This book consists of four chapters. The first chapter deals with the limits of the contemporary economic model. The first section presents the foundations of the economic system, namely the industrial revolution and the capitalist process. We will see that observers quickly began to focus on the environmental consequences of transformations in the field of production. The market economy develops with capitalism. The production and the mass consumption become the engine of the economic activity on a big part of the territories in the world and thus the source of a pollution. In the second section, we will analyze the consequences of this pollution by distinguishing between solid waste and pollutant emissions. We will present some figures, and then the main public policies to fight pollution, including the most recent laws in favor of the circular economy. Indeed, the State is a major actor in the ecological transition to encourage and also force companies to change their production methods. The third section will raise the question of the sustainability of our economic model. We will return to the distinction between strong and weak sustainability. The indicator of the ecological footprint will allow us to show how our economic model goes beyond the limits of the planet.

In the second chapter, we analyze eco-design as a source of environmental innovation and therefore an essential step in the ecological transition. In the first section, we present the precursors of eco-design, then the two visions of eco-design: on the one hand, technocentric eco-design, in which it is the improvements made possible by the technique that lead to the reduction of environmental impacts; on the other hand, sustainable design, in which the environmental and social dimensions are associated. In the second section, we will analyze technocentric eco-design more precisely, the most widely used method. Finally, in the third section, we will see how eco-design contributes to environmental innovations such as the circular economy, the functional economy and frugal innovation.

The third chapter proposes evaluating the contribution of eco-design approaches to the ecological transition. We will analyze eco-design practices in a systemic approach, in order to evaluate to what extent we can consider that there are factors favorable to the ecological transition and what type of eco-design is encouraged. We will study the characteristics of the socio-technical systems in which eco-design approaches have been developed. Thus, we will distinguish the activities of the actors (companies, states, training centers, users), the relationships they develop and the institutions that guide their strategies. We will conclude by studying the financial results and communication strategies. Indeed, eco-design approaches seem to be validated by the market: they respond to normative institutions, which support market values in socio-technical systems. As for communication strategies, they encourage eco-design approaches, most often based on life cycle analyses.

The objective of the last chapter is to analyze how eco-design can be a sustainable response to the environmental degradation resulting from the economic model and thus contribute significantly to the ecological transition. In the first section, we will note the limits of technocentric eco-design. It does not make it possible to escape from the dominant socio-technical regime. In addition, the search for eco-efficiency and the rebound effect prevent us from getting out of the ecological crisis. The second section is an analysis of the circular economy as an opportunity for new economic models, to which technocentric eco-design contributes. Nevertheless, it remains a partial solution, well identified by the principles of thermodynamics. In the third section, we will see what avenues can be envisaged in a sustainable design approach. The search for new uses with re-examined relations between the individual and the environment and frugal innovation are approaches that transform our relationship with objects. In the last section, we will study the offered perspectives by the processes of open innovations, allowing a greater participation of the actors and also the contributions of the notion of governance linked to the concept of common good. We will thus question the capacity of actors to negotiate, and even to establish new rules in order to create the conditions for strong sustainability.

1

The Environmental Impact of the Contemporary Economic Model

The report published by the IPCC (IPCC 2021) in August 2021 confirms the colossal impact of our economic model on the environment: increase in greenhouse gas emissions, surface temperature, precipitation, salinity at the surface of the oceans, surface water temperature and the retreat of glaciers, are most certainly linked to human activity.

The purpose of this first chapter is to identify the limits of our economic model in order to better define the objectives and contents of the ecological transition. Based on the three-part model of buy, consume and discard, the current economic model can lead to several dead ends: climate change disrupting our living conditions and provoking conflicts in certain territories, increasing difficulties in finding resources to produce and meet our needs as consumers, pollution causing health problems among populations. The ecological transition designates the transition from the current economic model to sustainable production and consumption methods. It then consists of finding the technical and human means that will probably change our lifestyles.

Before presenting the solutions developed by companies in the following chapters, we will look at the foundations of the economic model: the Industrial Revolution and the capitalist process. In the second section, we will study the most visible aspects of the limits of the economic model: the production of waste and emissions of pollutants. We will conclude with a reflection on the sustainability of the economic model.

1.1. The negative effects of the economic model on the environment

Two major changes explain the negative consequences of human activity on the environment: the Industrial Revolution from the middle of the 18th century and the capitalist system.

1.1.1. *Disasters of economic activity accelerated by the First Industrial Revolution*

The First Industrial Revolution started in England around 1740–1760, then spread to Belgium, France, Switzerland and later to Germany between 1770 and 1880 (Bairoch 1997a). It led to profound social transformations. A part of the population was literally put to work in factories in often very harsh conditions. They learned to work to the rhythm of the machines and in extreme noise. Long working days, accidents, sometimes fatal, and child labor were characteristic of the Industrial Revolution (Beaud 2010). The industrialization of the United States began a little later. It was driven by the massive immigration of certain European populations and the arrival of European capital. The lifestyles of the populations changed.

With the Industrial Revolution, the production of goods required the massive use of fossil fuels and mineral energies. They provided fire and heat, which played a decisive role in the revolution described by Jacques Grinevald as "thermo-industrial" (Bourg and Rayssac 2006). Coal was the main fossil fuel in England until the middle of the 20th century (Bourg and Rayssac 2006).

Industrial production was rapidly growing. At the beginning of the 20th century, the United States surpassed European production volumes (Table 1.1).

	1800	1830	1860	1880	1900	1913
Germany	8	9	15	25	52	85
United Kingdom	16	25	64	87	100	115
France	9	12	20	28	39	59
United States	9	14	21	38	69	126

Table 1.1. *Manufacturing output per capita (base UK 1900 = 100) (from Bairoch (1997a, p. 404))*

The negative consequences of the Industrial Revolution on the environment were quickly perceptible. Thus, in cities undergoing industrialization, questions were raised about air pollution from fumes resulting from the widespread use of steam-powered machines (Bonneuil and Frioux 2013) in the 19th century. In the 20th century, congresses were organized on the issue of air pollution (Frioux 2013) by municipalities.

What did economists say? During the Enlightenment period in the 18th century, some philosophers believed in the virtues of scientific and technical progress. It allowed individuals to achieve happiness, freedom and knowledge. But this period of economic, social and demographic change was a source of questioning (Veyret 2016). Philosophers and economists, such as Malthus, questioned the strong growth of the population while that of resources was slower. For Blaug (1985), Malthus (1798) showed in his *An Essay on the Principle of Population* that land scarcity was a limiting factor for economic progress. Veyret (2016) cites the German geographer Ratzel (1891), who questions the limits of resource use and points out the economics of looting.

During the 19th century, the ecological movement was constituted on the idea to protect nature. The fumes and pollution associated with rapid industrialization and urbanization gave rise to concern. The first nature protection movement, the Sierra Club, was created in 1892 in the United States by John Muir. In Europe, conferences were organized for the protection of Rhine salmon or birds of the African territories. In 1923, the director of the Muséum national d'histoire naturelle in France, Louis Mangin, denounced "the disastrous destruction that affects nature" at the first international congress on the protection of nature (Veyret 2016, p. 15).

The environmental issue has thus occupied an increasing place in the social representations during the 20th century. After World War II, the first controversies on the consequences of industrial activity on the environment appeared (Pessis et al. 2013). Céline Pessis reminds us that writers and scientists such as Reher Heim, Théodore Monod and Jean Rostand alerted us to the deterioration of the environment. Three major elements explain the general public's awareness of the risks associated with industrial activity:

– The publication in 1972 of the Meadows report, entitled "Limits to Growth". It was commissioned by the Club of Rome, a think tank bringing together scientists, economists, national and international civil servants as well as industrialists. It challenges the dominant idea of infinite growth and concludes that growth is limited in a finite world. This study is pessimistic about the possibilities offered by technical progress. It is based on an approach where, "whatever the way in which societies modify their use of technology, the population will in any case have an effect on the environment" (Debref 2018, p. 7).

– The holding of the first world conference on the environment in Stockholm: organized by the United Nations in 1972, it showed the awareness at the level of the States of the questions relating to the environment. The concept of ecodevelopment was introduced by the economist Ignacy Sachs. It is a question of considering social, environmental and economic aspects jointly. Social objectives were given priority, followed by environmental issues, then the search for economic viability (Figuière et al. 2014). The Stockholm conference led to the creation of the United Nations Environment Programme (UNEP), a complement to the United Nations Development Programme (UNDP).

– Major incidents linked to industrial activity because ecological disasters were becoming more frequent and causing more damage: sinking of the oil tankers Torrey Canyon in 1968 and Amoco Cadiz in 1978, explosion of a chemical plant in Bhopal in 1984, pollution in the chemical plant in Seveso and nuclear accident in Chernobyl in 1986.

The faith in progress of the Enlightenment philosophy was questioned: new technologies could lead to the worst disasters.

Today, the cumulative impacts of climate change, degradation of ecosystems and the depletion of natural capital on the global environment are recognized by the majority of society. The economic literature, for its part, has taken up the environmental issue since the 1960s (Godard 2015; De Jesus and Mendonça 2018).

1.1.2. *Questions about the capitalist system*

Our economic system is dominated by capitalism in most countries. Capitalism is characterized by the importance of private ownership of the means of production and the accumulation of capital, guided by the search for profit and the real driving force of economic growth (Plihon 2007). Capitalism is linked to the market economy. Economic activities are driven by the market, which is considered as the place where economic actors coordinate their activities. Progressively, market exchanges have a determining impact on relations between human beings.

Capitalism has experienced three stages: commercial capitalism until the end of the 18th century, industrial capitalism in the 18th century and, since the end of the 19th century, the capitalism of large industrial and financial groups (Plihon 2007). Commercial capitalism was based on the new trade routes with the New World and the great technical discoveries (printing). Industrial capitalism was born with the Industrial Revolution. It has transformed, among other things, production systems, social classes, mentalities and values, modes of existence and consumption, transport and communications (Beaud 2010). It is at the heart of the dynamics responsible for environmental degradation. Finally, the capitalism of large industrial and financial groups is accompanied by the globalization process. Global markets for goods, services and capital are emerging (Bocquet 2018). It is also accompanied by the development of the consumption society based on three verbs: buy, consume and discard.

Capitalism and the market are linked. Capitalism is a mode of governance with considerable transformative power in the social and environmental domains (Polanyi 1983; Beaud 2010). It is accompanied by the development of the market economy, which profoundly modifies social relations: “Society [becomes] in every respect an appendage of the economic system” (Polanyi 1983, p. 111). The prices of goods, labor, land and money are coordinated on the market. The individual is no longer motivated by the search for subsistence, but by the search for gain. Finally, both land and individuals are bought as raw materials. This marketization of society is very present today: the individuals are salaried employees of companies; natural resources, such as forests, are owned by companies or states in order to transform them into arable land or extract rare metals. In other words, “everything can be bought and sold” on the market. The concepts of ecodevelopment and sustainable development challenge our priorities: should we prioritize economic profitability, the preservation of resources, the well-being of individuals? We will return to this at the end of the chapter.

Consumption, a component of the market economy, reflects these social transformations, which as we know it today emerged in Europe around 1750. Marie-Emmanuelle Chessel (2012) explains it by the greater purchasing power of households, the increased diffusion of luxury and semi-luxury goods, the rise in the number of retailers in cities, advertisements in the press and the global circulation of goods. Mass markets, industrialization and cultural attitudes have allowed higher incomes to be used for ever-increasing production (Goodwin et al. 2018). Consumption accelerated during the 20th century to become mass consumption in the United States during the interwar period and then in Europe in the 1950s. With mass consumption, consumerist values and attitudes gained momentum. A few figures illustrate our point. Between 1914 and 1939, the number of motor cars in circulation increased from 2.5 to 50 million. Between 1922 and 1925 – when radio sets first appeared – and 1938, the number of devices increased from 0 to 97 million in the world. In 1938 in developed countries alone, nearly one in two households owned a radio set (Bairoch 1997c, pp. 44–46).

The model of mass consumption is linked to that of mass production, emblematic of the economies of the industrialized countries since the end of World War II. The Fordist system expanded with the mass production of manufactured goods, the scientific organization of work, large companies hiring thousands of employees, the rise of consumption of processed products enabled by regular wages (Postel and Rousseau 2008; Vaileanu-Paun and Boutillier 2012). The economic crisis in 1970–1980 was that of this model of industrial organization.

The spread of mass consumption and production is visible all over the world. Most countries, regardless of their per capita income or degree of industrialization, aspire to a mode of consumption and production similar to those of the so-called developed countries. The exchange of goods, capital and services is globalized. Bocquet (2018, p. 18) emphasizes the "interdependence between all the elements (societies, nations, companies, cultures) that constitute a complex system". Economic growth is thus imposed on all societies, from the richest to the poorest, and it is accompanied by a strengthening of inequalities between individuals (Beaud 2010). This economic model is becoming more widespread, but what is the capacity of the planet to absorb the growing extraction of resources necessary for production, the increasing pollution of the environment and the increasing pollution of air, soil and water, or the generation of waste that harms individuals and living beings? We will return to this question in section 1.3.

We have studied the Industrial Revolution and capitalism. These processes are linked and have led to mass production and consumption which implies increasing pollution. We will now look at solid waste and pollutant emissions.

1.2. A society that pollutes

The impact of our society on the planet is summarized by Bihouix (2014) as follows:

> I buy a cell phone in France, and by doing so I have exploited miners in the Congo, destroyed forests in Papua, enriched Russian oligarchs, polluted Chinese water tables, and then, twelve to eighteen months later, I will dump my electronic waste in Ghana or elsewhere.

This short quote highlights several forms of pollution. Our lifestyles generate solid waste as well as polluting emissions to water and air. We are interested in waste: a resource lost to its owner, by individual or company, as well as to the rest of society in the absence of recycling or recovery. Thus, our lifestyles do not allow us to give a value to all the resources used in production. Their accumulation, most often on the outskirts of cities, is symptomatic of our society. More recently, the growth of digital activities also generates waste, not always visible to users.

We will start by defining the concept of waste, then we will present some statistical data. We will conclude with a presentation of the main public policies concerning waste.

1.2.1. *Waste: elements of definition*

Waste is a fairly recent phenomenon in our history. Until the end of the 19th century, in French cities, the economy of recycling made it possible to have no "waste". Materials from recycling were re-injected into the economic process and allowed other products to be made (Aurez and Georgeault 2016a, p. 40). Companies were creating material loops long before the concept of circular economy was coined (Gallaud and Laperche 2016; Korhonen et al. 2018a). Since the beginning of the 20th century, waste literally proliferated, and from 1945 onwards, landfill became the normal way of managing waste (Pessis 2013).

Waste is a material or object that is thrown away because it is no longer needed or has no value. More precisely, the term waste has two meanings from an etymological point of view. It refers to an object that has lost its value "over time, with wear and tear and decomposition of its attributes, or over space because it does not have the same utility everywhere" (Lupton 2011, p. 23). This loss of value results in waste. Waste is also a by-product from manufacturing processes (Lupton 2011).

The French and European legal definitions of waste are relatively close. According to the law of July 15, 1975, this is considered as constituting waste:

> Any residue of a production, process of production, transformation or use, any substance, material, product, or more generally any movable property abandoned or that the holder intends to abandon (Article L. 541-1-1 of the Environmental Code).

According to the European Union, waste is defined as "any substance or object [...] which the holder discards or intends or is required to discard" (Directive 2006/12/EC of the European Parliament and of the Council of April 5, 2006 on waste).

The issue of waste can also be approached from an economic, environmental and sociological point of view.

In an economic approach, so-called waste products are unwanted, fatal productions of the production or consumption activity of an economic agent, for which it will have to pay for removal, recycling and disposal services (Bayon and Levratto 2009). They thus pose an economic problem, which is that of the costs of waste management borne by businesses, households and communities. At the same time, the presence of waste reflects an under-utilization and waste of available resources, reducing the efficiency of the use of resource sources. Finally, waste is a loss of value (Lupton 2011), a "flow that is not profitable for the company" to quote ADEME (ADEME 2014).

From an environmental point of view, waste, as well as transportation and material extraction, are a source of pollution. Their impact can be temporary (biodegradable waste) or permanently negative, for example, plastic waste does not degrade and accumulates in the environment.

Finally, in a sociological approach, waste is the result of a "process of abandonment by which the object passes from ownership to inadequacy, and thus loses its use value, if not its exchange value" (Corteel and Le Lay 2011). Like the economic application, the loss of use value and perhaps of exchange value raises questions for the manager concerned with controlling the company's resources.

1.2.2. *Pollution in the world: an inevitable growth?*

Our activities contribute to air, water and soil pollution. Solid waste is the most visible form, but emissions of pollutants into the air, fresh water, oceans, land and soil and pollution from chemicals also have a major environmental impact.

1.2.2.1. *Solid waste*

We have data on the production and treatment of waste at three levels of analysis: global, European and French. At this last level, we will see that the main producer is the building and public work sector.

Solid waste generation in the world varies greatly by region, income level and to a lesser extent by urbanization rate. For example, waste generation per person per day ranges from 0.52 kg in South Asia to 1.18 kg in Europe and Central Asia and 2.21 kg in North America according to World Bank data (Table 1.2). The total volume of waste is projected to increase from 2.01 billion tons in 2016, to 2.59 in 2030, to 3.40 in 2050 (Silpa et al. 2018).

Looking at waste processing patterns, there is a wide variation in collection rates, from 44% in Sub-Saharan Africa to 96% in Europe and Central Asia and 100% in North America (Table 1.3). More than half of the treated waste is sent to landfill in North America and Latin America and the Caribbean. In other words, the waste is left in the open air with the known consequences: deep water and surface water pollution, as well as pollutants for local residents (odors, gases, smoke, etc.). The highest recycling rate is found in North America, Europe and Central Asia.

	Millions of tons per capita	kg/person/day
North America	289	2.21
South Asia	334	0.52
Middle East and North Africa	129	0.81
Europe and Central Asia	392	1.18
Sub-Saharan Africa	174	0.46
Latin America and the Caribbean	231	0.99
East Asia and Pacific	468	0.56

Table 1.2. *Global waste generation by region in 2016 (from Silpa et al. (2018))*

	Waste collection rate	Discharge	Incineration	Recycling	Other treatments
North America	100%	54.3%	12%	33.3%	12.3%
South Asia	44%	4%	–	5%	91.0%
Middle East and North Africa	82%	34%	> 1%	9%	56.0%
Europe and Central Asia	90%	25.9%	17.8%	20%	36.3%
Sub-Saharan Africa	44%	24%	–	6.6%	69.4%
Latin America and Caribbean	84%	68.5%	–	4.5%	27.0%
East Asia and Pacific	71%	46%	24%	9%	21.0%

Table 1.3. *Global waste treatment by region in 2016 (from World Bank (2018))*

Among solid waste, we must distinguish electronic waste. The rapid growth of digital technologies has accelerated its production while the recycling rate remains low. It contains materials that can either be reused or recycled or can be recycled into other materials. The Global E-waste Monitor report, published in partnership with the

United Nations Institute for Training and Research (UNITAR) (Forti et al. 2020), provides very interesting data. In 2019, 53.6 million tons of e-waste were generated worldwide and 74 million tons are expected in 2030, a projected increase of 38%. Of this total, the collection and recycling rate is 17.4% (Table 1.4).

	Electronic waste generated (millions of tons)	Electronic waste generated (kg/capita)	Electronic waste collected (millions of tons)	Share of collected waste recycled
Africa	2.9	2.5	0.03	0.09%
America	13.1	13.3	1.2	9.4%
Asia	24.9	5.6	2.9	11.7%
Europe	12	16.2	5.1	42.5%
Oceania	0.7	16.1	0.06	8.8%

Table 1.4. *E-waste: generation, collection and recycling by continent in 2016 (from Forti et al. (2020))*

	2004	2006	2008	2010	2012	2014	2016	2018
Total waste treated (millions of tons)	1,827	1,858	1,877	1,959	2,039	2,039	2,022	2,087
Discharge	50.8%	45.1%	44.1%	42.4%	43.6%	42.2%	40.5%	38.7%
Incineration	1.8%	2.1%	2.1%	1.6%	1.3%	1.0%	0.7%	0.5%
Recycling and backfilling	42.6%	47.9%	48.7%	47.0%	45.7%	46.1%	47.2%	49.1%
Valuation energy	3.3%	3.6%	4.1%	4.2%	4.6%	5.0%	5.7%	5.9%
Other	1.5%	1.3%	1%	4.8%	4.9%	5.7%	5.9%	5.9%

Table 1.5. *Non-hazardous waste treatment in the European Union, 2004–2018 (based on Eurostat (2022))*[1]

For Europe, the total volume of waste increased from 1,897 to 2,169 million tons between 2004 and 2018 according to Eurostat data

1 See: https://ec.europa.eu/.

(2022)[2], a growth of 14.3%, while over the same period the European population grew by 2.9% (Table 1.5). The non-hazardous waste makes up 96% of total waste. As regards the treatment of non-hazardous waste, there has been a decrease of almost 12 percentage points in the percentage of waste going to landfill, offset in particular by the growth in recycling and energy recovery.

For France, the total volume of waste increased by an equal amount. Similarly, there has been a decrease in the share of waste going to landfill (6.4 percentage points), which can be attributed mainly to increased recycling (Table 1.6).

	2004	**2006**	**2008**	**2010**	**2012**	**2014**	**2016**	**2018**
Total waste treated (millions of tons)	277	286	316	323	306	292	298	316
Discharge	32.9%	33.7%	33.2%	33.2%	33.1%	28.8%	27.2%	26.5%
Incineration	0.3%	2.1%	2.3%	2.0%	1.9%	1.6%	1.2%	0.8%
Recycling and backfilling	62.0%	60.8%	60.8%	61.2%	61.3%	65.1%	66.2%	67.1%
Energy valorization	4.6%	3.1%	3.5%	3.4%	3.4%	4.2%	5.2%	5.3%
Other	0.2%	0.4%	0.2%	0.2%	0.3%	0.3%	0.2%	0.2%

Table 1.6. *Non-hazardous waste treatment in France, 2004–2018 (based on Eurostat (2022))*[3]

In France, the data provided by ADEME allow us to distinguish waste according to its origin (Table 1.7). The construction sector is the largest producer of waste, ahead of businesses (four times more in kg) and households (nearly five times more). Waste collection from households (excluding separate collections and waste collection

2 See: https://ec.europa.eu/.

3 See: https://ec.europa.eu/.

centers)[4] doubled between 1960 and 2006 to 354 kg/person/year (ADEME 2009), then decreased to 249 kg/person/year in 2019 (ADEME 2019).

	Household and similar waste[5]	Businesses (excluding construction and similar)	Construction	Total
kg/capita	580	900	3,400	4 880
kg/capita (construction sector = 100)	17	26	100	–
Millions of tons	32	65	224	321

Table 1.7. *Origin of waste in France in 2017 (according to ADEME (2020b) and our own calculations)*

1.2.2.2. *Pollutant emissions: the example of air pollution*

Finally, in addition to solid and electronic waste, our production and consumption patterns generate other forms of pollution.

To understand the issues of pollutant emissions, we will take the example of air pollution. According to ADEME, transport, heating of buildings, agriculture, industry, energy production and open burning of waste are the main activities that cause air pollution (Table 1.8).

The pollutants generated are numerous (Box 1.1). Their concentration in the atmosphere contributes to climate change. They have harmful effects on health. For example, air pollution with fine particles is responsible for 48,000 deaths each year in France (ADEME 2020a).

4 Other household wastes include separate collections of glass, dry recyclables, bio-waste, bulky items, hazardous waste and waste collection centers.

5 Household and similar waste refers to the waste collected by municipalities from households, craftsmen, shops and small businesses.

	Transport	Residential	Agriculture	Industry
Nitrogen oxides	63% diesel road transport (cars and trucks)	11% wood heating	7% tractors, heating of greenhouses	19% manufacturing industry (building materials, chemicals)
PM_{10}	15% diesel road transport and wear dust	34% wood heating, green waste fires	24% crops (ploughing), livestock buildings	27% construction, industry
Volatile organic compounds (VOCs)	11% road transport gasoline and waterway	45% wood heating and domestic solvent use	2% machines, engines, heating of greenhouses	42% construction, chemical industry, food industry
Sulfur dioxide (SO_2)	2%	18% heating with domestic oil	1%	79% oil refining, building materials, metallurgy, chemicals, thermal power plants
Ammonia (NH_3)	1%	3%	94% breeding, spreading, mineral fertilization	2%

Table 1.8. *Main air pollutants and their main sources (from ADEME (2020a))*

Particles are complex pollutants differentiated by their diameter, less than 10 micrometers (10 thousandths of a millimeter) for PM_{10} and less than 2.5 micrometers for $PM_{2.5}$. Their toxicity is related to their chemical nature and size. The finest are the most dangerous to health, as they easily and durably penetrate the organism.

Particles can be primary pollutants when they are directly emitted into the atmosphere, or secondary when they are generated from other pollutants known as "precursors".

Beyond particulate matter, the main pollutants are:

– certain gases such as nitrogen oxides (NO_X), tropospheric ozone (O_3), ammoniac (NH_3), sulfur dioxide (SO_2), carbon monoxide (CO);

– volatile organic compounds (VOCs): benzene, formaldehyde, isoprene, etc.;

– persistent organic pollutants (POPs), including polycyclic aromatic hydrocarbons (PAHs);

– heavy metals: lead, mercury, cadmium, nickel, etc.;

– biological pollutants: legionella, pollens, molds.

Box 1.1. *The different air pollutants (from ADEME (2020a, p. 7))*

Statistics provided by *Our World in Data* (an online publication run by Oxford University) give us an international view of air pollutant emissions per person (Table 1.9).

Oceania (which includes Australia and New Zealand) and North America are the main emitters. Australia is particularly known for its massive use of fossil fuels with a large mining sector that accounts for sulfur dioxide emissions (*Le Monde*, November 20, 2015).

	Nitrogen oxides NO_X (kg)		Non-methane volatile organic compounds NMVOCs (kg)		Sulfur dioxide SO_2 (kg)		Ammonia NH_3 (kg)	
	1750	2019	1750	2019	1750	2019	1750	2019
Africa	0.23	5.19	5.31	20.64	0.10	4.93	4.10	6.47
North America	0.52	20.98	22.16	28.61	0.33	10.15	4.64	9.77
South America	0.39	15.08	6.86	19.75	0.13	9.38	5.62	12.47
Asia	0.17	12.84	7.23	16.97	0.20	9.93	4.03	7.50
Europe	0.50	14.88	17.85	16.10	1.44	11.06	5.57	8.82
Oceania	0.13	35.67	1.64	27.09	0.03	31.73	2	19.64

Table 1.9. *Major air pollutants by continent per capita (from Our World in Data (2022)[6])*

1.2.3. *Public policies to reduce pollution*

The State is a major player in the ecological transition to encourage and also force companies to change their production methods. Public policies were rather regulatory until the beginning of the 1990s; other instruments have been developed since 1992 with environmental taxes, technological standards and tradable emission permits (e.g. Lupton 2011). Companies working in Europe are progressively being asked to reduce pollution, especially waste, at the source.

The commitments made by Europe at the Earth Summit in Rio de Janeiro in 1992 marked a turning point in European policy on waste management and prevention of waste. The European Council then adopted the first sustainable development strategy of the European Union. The European policy will gradually introduce the principle of extended producer responsibility, or "polluter-pays" principle. This principle, introduced by the OECD in 1972, aims in particular to limit the production of goods and waste by making producers responsible

6 See: https://ourworldindata.org/.

for managing the end-of-life of products (Godard 2015). They must thus bear the costs of waste removal and treatment. This is a concept inspired by Pigou's approach in which waste is taxed: the manufacturer pays for the removal of waste which it produces instead of communities (Boutillier 2003; Tran Dieu and Vernier 2017). He is responsible for the recovery and treatment of end-of-life products through specialized channels. The extended responsibility of the producer is carried out collectively through eco-organizations. This policy has been visible in France since 1994 (Box 1.2). In 2001, the European Commission published the Green Paper on Integrated Product Policy. It then truly instituted the polluter-pays principle, which will be implemented by taxing the most polluting products.

In France, the most well-known eco-organization is Eco-Emballages, created in 1994 and now called Citeo. It finances the selective collection and recycling of packaging waste on behalf of large-scale consumption companies. Also in France, since 1999, companies are encouraged to reduce waste, notably through the TGAP (general tax on polluting activities).

Since 2007, the Grenelle de l'environnement has been strengthening public policies in favor of the environment and sustainable development. We can mention the expansion of the categories of waste concerned by the extended producer responsibility, the generalization of sorting signs on products for consumers and the modification of the regulations on the TGAP with a new increase and the creation of new components (on paints and solvents, for example).

Box 1.2. *Extended producer responsibility in France and the Grenelle agreements*

In 2001, Europe went further and launched the integrated product policy. The objective is to promote the launch onto the market of ecological products through a range of tools. This incentive policy has three components: taxation through the internalization of external environmental costs, the application of the polluter-pays principle and the producer's environmental responsibility for the product manufactured (Le Pochat 2005; Pôle éco-conception 2022a).

Finally, in 2008, Europe replaced its waste management policy, which was aimed at curating waste and consisted of building treatment centers, with a waste prevention policy consisting of reducing waste at the source. The new 2008 directive (Commission européenne 2008) aims to minimize waste upstream of its production. Companies must develop strategies to prevent a substance, material or product from becoming waste. The directive establishes a waste hierarchy according to which Member States must take measures for the treatment of their waste. The order of priority is as follows:

– prevent waste;

– prepare waste for reuse;

– recycle waste;

– waste recovery, in particular through energy;

– dispose of the waste if none of the above actions are possible.

The 2008 directive proposes new definitions of waste, by-products, and also criteria for ending waste status and the need to choose appropriate technologies (Berneman and Vernier 2011). Setting the boundary between waste and by-products is strategic. A by-product is a substance or object resulting from a production process whose primary purpose is not the production of the said good. Under certain conditions, it is no longer considered a waste: its subsequent use is then legal and does not require any additional treatment; and is no longer subject to waste taxes. The by-product must also meet environmental and health requirements. The criteria for determining whether a product can be recycled are also decisive, as they determine how it is treated: reuse, recycling, recovery or disposal (Box 1.3).

Policies to reduce waste and environmental pollution are also designed by product type (see Table 1.10). They require companies to change their strategies to reduce waste in specific areas. In France, we should also note the 2023 food waste law, aimed at combating food waste.

Recycling is any operation whereby waste materials are reprocessed into products, materials or substances for their final function or other purposes.

Recovery is based on reuse, recycling or any other action aimed at obtaining reusable materials or energy from waste.

Incineration is a form of recovery. This method of treating waste consists of combustion (technology and temperature vary according to the nature of the waste) and treatment of flue gases. This technique produces three categories of waste: slag, ashes and flue gas cleaning residues. The heat produced by the incineration process is used for energy recovery (production of electricity and heat) in most units. In the case of fermentable waste (biodegradable organic matter), the recovery consists of either methanization or composting the waste.

Disposal or burial is the oldest method of waste destruction, which consists of burying it underground on finished sites. These sites are now called CSDUs (*centers de stockage des déchets ultimes*), because, according to the regulations, only waste that can no longer be recovered in acceptable technical and economic conditions is admitted.

Box 1.3. *Waste treatment (from Berneman and Vernier (2011; ADEME 2020b))*

Since the late 2000s, some states have introduced legislation to support the circular economy. They are now the most advanced form of public policies to fight against waste production. For example, China in 2009, and then Europe in 2015, have placed the circular economy at the heart of their policies with the aim of drastically reducing waste. The circular economy is based on the principle of material and energy loops in order to produce and then consume by minimizing the extraction of materials and maximizing their reuse.

We will present the principles of the circular economy in more detail in the next chapter.

Field of application	Description and objective
Packaging (1994)	The directive aims to prevent the formation of packaging waste (recovery and recycling) and establishes the implementation of systems for the return and collection of packaging.
End-of-life vehicles (2000)	The end-of-life vehicles directive (2000) aims to recover, recycle and transform car wrecks.
Electrical and electronic equipment (2005)	The WEEE directive (waste electrical and electronic equipment, 2005) makes producers of electrical and electronic equipment responsible for their products at the end of their life. They must contribute to the financing of their collection, treatment, recovery and disposal (Commission européenne 2005).
Substances deemed hazardous to the environment (2006, 2007)	The European directive RoHS (Restriction on the use of certain Hazardous Substances, 2006) on electrical and electronic equipment establishes a list of prohibited substances (Commission européenne 2007). The REACH regulation (Registration, Evaluation, Authorization and Restriction of Chemical Substances, 2007) considers that the producer is responsible for managing chemical risks and must provide health information on these products (Commission européenne 2007).
Eco-design (2009)	The European eco-design directive aims to gradually remove the least efficient products from the market from an environmental point of view and in the energy field as a whole (Commission européenne 2009).

Table 1.10. *Examples of European waste directives by the type of product*

The Law for the Promotion of Circular Economy of the People's Republic of China came into effect on January 1, 2009. It is the first law making the circular economy a national strategy (Kirchherr et al. 2018). It proposes incentive measures and more specific measures such as the development of pilot industrial parks, investment funds for projects dealing with the circular economy and reduced VAT on ecoproducts (Aurez and Georgeault 2016b).

In Europe, the policy in favor of the circular economy is based on actions that encourage recycling and the prevention of recoverable material losses, job creation and growth, design of new economic models, eco-design, industrial symbioses, and reduction of greenhouse gas emissions and environmental impacts (Bocken et al. 2016; Cainelli et al. 2020). As we will see below, designing the circular economy while demanding continued economic growth is the subject of debate between proponents of strong sustainability and those of weak sustainability.

In 2020, the European Commission adopted a second action plan for a circular economy, including the "Green Pact for Europe", with the aim of overcoming the Covid-19 pandemic and reviving the economy[7]. Actions are being taken to design more sustainable products that can be repaired, recycled and reused[8]. The scope of the eco-design directive is extended to non-energy-related products.

1.3. A sustainable economic model?

Environmental imbalances and social inequalities are undeniable. As previously mentioned, they originate from the socio-economic system that characterizes our society based on market capitalism and industrialization since the middle of the 18th century (Beaud 2010; Buclet 2011).

We are in the era of the Anthropocene, where the effects of human activities account for the profound degradation of the "fragile physico-chemical balance that has made life on Earth possible" (Beaud 2010, p. 389). The notion of sustainability and the ecological footprint indicator make it possible to analyze how our production and consumption patterns lead to exceeding the planet's capacity to

7 https://www.europarl.europa.eu/news/fr/headlines/society/20210128STO96607/comment-parvenir-a-une-economie-circulaire-d-ici-2050.
8 https://ec.europa.eu/info/strategy/priorities-2019-2024/european-green-deal_fr.

provide the resources necessary for the functioning of the economic model.

1.3.1. *The sustainability approach*

The concept of sustainable development is a proposed response to these challenges. It succeeds the term ecodevelopment, introduced by Ignacio Sachs in 1972 at the Stockholm conference. According to the report "Our Common Future" (Brundtland 1987), the objective of sustainable development is to meet the needs of present generations while allowing future generations to meet theirs. In this perspective, the objective of our societies is to achieve human well-being (Buclet 2011) by considering the imbalances in the environment and social inequalities. Sustainable development is therefore the search for a development that simultaneously associates the economic, social and environmental dimensions.

The term sustainability allows us to question the capacity of our society to meet the needs of future generations. It questions the forms taken by contemporary global capitalism. The literature in economics studies it in two ways: weak sustainability and strong sustainability.

From a weak sustainability perspective, human-made capital (created by wealth) and natural capital (natural resources) are substitutable. According to neoclassical economists, technical progress allows this substitution. It must be possible to transmit a capital stock from one generation to the next whose composition will vary in order to allow future generations to meet their need for well-being (Vivien et al. 2013). Thus, "current generations, which consume natural capital must bequeath more human-created capital to future generations" (Vivien et al. 2019, p. 31).

From a strong sustainability perspective, artificial capital and natural capital are not substitutable (Figuière et al. 2014; Vivien et al. 2019). Therefore, a stock of natural capital must be able to be handed down to future generations. This is the analysis held by ecological economists. It must be reconciled with the thermodynamic approach according to which the physical limits of the biosphere work against limitless economic growth (Vivien 2003; Bürgenmeier 2008; Froger et al. 2016) (Box 1.4).

According to the laws of thermodynamics, the functioning of the economy is analyzed as a system that depends on the biosphere. The flows of energy and matter are studied by considering the planet as a closed system. Thermodynamics is based on two principles. First principle: the mass of matter and energy remains constant. Second principle: energy is used to do work. According to the phenomenon of entropy, a part of the energy used for this work is transformed into heat. This phenomenon of dissipation in the form of heat makes the energy useless.

In the production process, according to the first principle, the human species cannot produce energy and matter; it can only transform them. The second principle explains an irreversible loss of energy: all work consumes energy which is dissipated in a production process. The disappearance of energy is progressive and inevitable.

In other words, our consumption-based lifestyle cannot continue. The economic process of transforming resources leads to an irreversible loss of energy. In the long run, growth would be limited.

Box 1.4. *The limits of the biosphere: the thermodynamic approach (from Vivien (2003), Bürgenmeier (2008) and Froger et al. (2016))*

The social dimension does not appear much in this strong/weak sustainability debate, as it is forgotten in favor of questions about growth and the environment (Clément et al. 2011). Some studies in economics are interested in social aspects with a reflection on the capacity to meet the needs of current and future generations (Vivien 2013). These analyses focus on four aspects:

– from the perspective of individual well-being and according to the approach of Amartya Sen, the capacities of individuals to achieve "states of being" such as moving, housing, being socially recognized;

– the notion of equity within and between generations;

– poverty and in particular the equal and inclusive access of the poorest to environmental negotiations;

– the content of growth, which differs according to the activity. Thus, the composition of growth differs according to whether it originates in the arms industry or in school construction.

We will thus define the social dimension of sustainable development by the search for the well-being of individuals, inter- and intra-generational equity, the reduction of poverty and the distribution of the gains from growth between activities. We define strong sustainability by closely associating the environmental and social dimensions (Clément et al. 2011). It is an axis of analysis in order to understand what the ecological transition can be.

1.3.2. *Ecological footprint*

We can complete the analysis of the concept of sustainability with the notion of the ecological footprint indicator. It shows that our society has exceeded the limits of the planet and is in a strong sustainability perspective. This indicator allows us to understand and measure the dependence of our economic and social system on the planet. According to Boutaud and Gondran (2018), the sphere of human activities (the econosphere) is intimately dependent on the biosphere from which it draws its energy and raw materials.

The authors compare two elements (Boutaud and Gondran 2018, p. 33):

1) The quantity of services from the biosphere needed to keep the econosphere functioning (the demand for nature).

2) The quantity of these same services that the biosphere is able to regenerate (the supply of nature).

In the case where (1) is greater than (2), then natural capital degrades.

However, the biosphere is constrained: consuming more resources than nature can generate is not possible in the long run. Boutaud and Gondran (2018) cite the case of the inhabitants of Easter Island. Their disappearance results, among other things, from the use of wood beyond the island's regeneration capacities. Their ecosystem has deteriorated, the wood has disappeared from the island and their civilization has collapsed.

The principle of the ecological footprint consists of "representing the ecosystem's quantity of regenerative capacity that must be mobilized to provide the econosphere with the resources it needs and assimilate the waste it produces. This regenerative part of the ecosystem therefore essentially concerns the biosphere" (Boutaud and Gondran 2018, p. 42).

It is measured by six indicators (Box 1.5).

1) Carbon footprint: area of forest theoretically required to sequester the part of CO_2 that is emitted into the atmosphere during the consumption of fossil fuels and which is not assimilated by the oceans.

2) Crops: areas needed to produce goods of vegetable origin consumed during the year (food and textile fibers).

3) Pasture: area of pasture necessary to satisfy the consumption of goods from livestock (meat, milk, wool, leather).

4) Forest footprint for wood and paper production.

5) Aquatic space to provide fisheries resources.

6) Bioproductive land use for urbanization and various infrastructures.

Box 1.5. *The six indicators for measuring ecological footprint (from Boutaud and Gondran (2018))*

According to the authors, the global ecological overshoot can be dated to the early 1970s. It results above all from the inability of the biosphere to assimilate the CO_2 released by the combustion of fossil fuels. The reduction in soil productivity (with in particular the physico-chemical degradation of the environment, erosion, the accumulation of toxic substances, pesticides) and the overexploitation of certain renewable resources, which lead to the reduction of certain stocks, for example, also explain this excess.

1.4. Conclusion

This first chapter shows that the current situation is the result of the changes initiated by the First Industrial Revolution. The capitalist system, by supporting production and consumption, has made them permanent. We therefore live in a world that generates pollution, particularly in the form of solid waste. The public authorities then developed public policies based on the management and then the prevention of waste. Recent laws in favor of the circular economy seem to be the most likely solutions to reduce pollution by eliminating or at least drastically reducing the generation of waste. Nevertheless, some companies voluntarily develop products that also reduce pollution as part of CSR (corporate social responsibility) strategies, as we will see in Chapter 3. We ended this chapter by questioning the sustainability of our society from an environmental point of view. This book only deals with the environmental dimensions of our economic model, choosing not to elaborate on social consequences, which are nonetheless very real and sometimes even disastrous.

The ecological transition consists of modifying the current economic model to move towards a more sustainable model, making it possible to achieve human well-being while taking into account environmental imbalances (finitude of resources and environmental footprint linked to production methods and consumption). Companies, through their ability to offer new products, processes, organizations or marketing methods, must participate in this change. The object of the following chapter is the study of eco-design as a tool contributing to this transition.

2

Eco-design: Definitions and Theoretical Outlines

We have noted the growth of pollution associated with industrialization and the capitalist mode of production. As a source of short-, medium- and long-term pollutants, it weakens the living conditions of current and future generations. Companies are capable of producing the goods and services of our daily lives in a different way: this is innovation. They can also produce with a reduced environmental impact compared to conventional production methods: this is environmental innovation. We propose going further. The purpose of this chapter is to analyze a design method that is a source of environmental innovation: eco-design, which aims to limit the environmental impact associated with a product from its design to its end-of-life.

In the first section, we will show that the term eco-design covers two main categories of practices: on the one hand, technocentric eco-design, where technology is privileged, and, on the other hand, sustainable design, which combines economic, environmental and social dimensions. In a second section, we will analyze the first form, which dominates in companies. Finally, in a third section, we will study the other forms of environmental innovation by defining the contribution of eco-design.

2.1. Eco-design: a plural concept

The concept of eco-design has evolved around two approaches: technocentric eco-design, according to which technology can reduce environmental impact, and sustainable design, based on the search for harmony between the human species and its environment. The first perspective dominates current definitions of eco-design and corporate practices with life cycle analysis (LCA).

2.1.1. *The pioneers of eco-design*

Eco-design is a design method that must be associated with design and engineering. For the designer, it involves designing tools and objects with which the individual transforms their environment, and also their person (Papanek 1974). The designer relies on "a process of observation of uses, research in relation to the existing, analysis, then creation" (Designers + 2014, p. 4). The creativity is thus decisive in the approach by the designer. Design connects the individual with the environment in the broadest sense. It is also at the heart of the consumer society synonymous with the production of mass-produced objects, which must be desired in order to be purchased. It is thus a "practice [which], in new objects, tries nothing less than to reconcile capitalism and humanism, interest and meaning, seduction and authenticity, rhetoric and truth, tradition and invention" (Hatchuel 2006, p. 2). Eco-design will develop essentially on the need to preserve the planet while designing or improving products and to a lesser extent on the necessity of meeting the needs of individuals.

Architect Richard Buckminster Fuller and designer Victor Papanek can be considered the pioneers of eco-design. Lanoie and Plouffe (2008) cite Buckminster Fuller and Mc Hale, who published two reports in 1963 and 1964 on the limits of natural resources (Buckminster Fuller and Mc Hale 1963, 1964). To avoid the depletion of certain resources, they propose the reuse of discarded materials and the extension of the life span of products. They also advocate the development of solar energy, given the non-renewable nature of fossil fuels (Lanoie and Plouffe 2008).

As for Victor Papanek, he opposes a vision of design in the service of marketing and defends a responsible design that is concerned with the needs of humanity (Abrassart and Aggeri 2002). Design must challenge an approach where too many products, energy and irreplaceable resources are needed to design products (Papanek 1974). Kazazian, a pioneer of eco-design in France and internationally, will take up his legacy. He coordinated a book defending a light and sustainable approach to objects with a society of use of services and no longer of consumption (Kazazian 2003).

In the 1980s, designers became closer to manufacturers. Madge (1997) points out that Evelyn Möller in 1982 invented the expression "ecological functionalism" and established an ecological checklist for product designers and manufacturers. Then, Paul Burall (head of publicity at the UK Design Council) and John Elkington (environmental consultant) organized the exhibition *The Green Designer* in 1986. The objective was to show that green design was not opposed to the industrial world (Madge 1997). It should be noted that it was John Elkington who introduced the notion of the triple bottom line a few years later, where the performance of a company must be evaluated according to the three pillars of sustainable development (Elkington 2013). For some designers, the environmental problem can no longer be ignored.

The development of eco-design approaches follows two trends: technocentric eco-design and sustainable design.

2.1.2. *Technocentric eco-design*

Eco-design, which we will call here technocentric, is defined by its purpose: minimizing the environmental impact of a product compared to other similar products on the market and over all stages of the life cycle.

2.1.2.1. *Issues and founding principles*

We are in a technocentric vision according to which science and technologies are able to manage the environment for the benefit of present and future generations. Progress, efficiency and control are at the heart of the analyses. Choices are then determined by economic and industrial objectives (Debref 2018). Technological improvement aims to reconcile product design and respect for the environment. In other words, technology allows us to "move to a green mode of production without challenging our lifestyles" (Petit 2015, p. 36). Petit (2015) refers to this approach as "design of the environment": the human species adapts to the environment through technology. It is therefore a technical solution to the question of sustainability. We can then identify the objects created by the human species as artifacts aiming to solve the problems posed by environmental limits. Technocentric eco-design is growing in industry and among engineers (Guilloux 2015; Petit 2015). The environment is an objective and quantifiable entity. The thinking then consists of reducing the gaps between projected and perceived qualities, using evaluation methods (Guillou 2015). It is based on the design of technical systems that must be controlled (Abrassard and Aggeri 2002). Eco-design is then seen as "the integration of the environment parameter into the product development process" (Janin 2000, p. 31; Roux et al. 2005).

From the 1960s onwards, the design process was rationalized. Authors rely on functional analysis in order to find, order, characterize and prioritize the functions associated with a need. The design aims "to make a compromise between the utility (the functionalities) of a product or service and the cost (the impacts) associated with this utility" (Patingre 2001, p. 31). The design sciences will then try to integrate all the parameters of the life cycle of the product to be designed. The objective is to contribute to the control of the effects on the environment linked to the production, use and disposal of products and services (Patingre 2001). These impacts are assessed using LCA.

Manufacturers will integrate the environmental parameter into their product development process with a view to offering products that are less harmful to the planet. Two methods will evolve in parallel with an analysis to apprehend the environmental impact adopting either a mono-criteria or a multi-criteria approach.

2.1.2.2. *History of single-criteria approaches*

Among the single-criteria approaches, Lanoie and Plouffe (2008) have identified three ways of integrating eco-design.

The first wave of environmental integration in product development is associated with the material flow loop approach. Lanoie and Plouffe (2008) cite the thesis of Millet (1995):

> This approach consists of assuming that the environmental impacts caused by the product will be minimized if the flow of materials occurring in the end-of-life phase of the product's life cycle is looped.

The "material flow loop" approach emerged during the 1970s in response to landfill saturation and complaints from residents of the sites. Similarly, the incineration of waste was a contentious issue. One solution was to reduce the volume of waste by trying to recover as much as possible of what could be reused, recycled, etc. The regulations were curative and based on objectives to reduce the volume of solid waste. Selective collection systems were developed to collect large volumes of waste that could be easily recycled (paper, glass and metals). Waste volumes have decreased and new processes have made it possible to make new recovery and recycling activities profitable.

Industry had seen the economic benefits of recycling. For example, the energy consumption required to manufacture certain materials from natural resources was much higher than the energy required to remelt or process recovered materials. In the case of aluminum, this difference was as high as 90%. Nevertheless, this did not reduce the volume of waste sufficiently.

In the 1970s, the chemical industry was considered responsible for air and water pollution. Under pressure from pro-environmental groups, laws were passed to stop these industries from discharging their waste directly into the natural environment. The costs associated with treating the waste to make it "cleaner" are so high, in some cases, that governments have provided funding to prevent the closure of too many companies (Lanoie and Plouffe 2008). We mentioned in the first chapter the adoption by the OECD of the polluter-pays principle. This is a concept inspired by Pigou's approach in which waste is taxed: manufacturers pay for the management of the waste they produce instead of authorities (Boutillier 2003). This polluter-pays principle will progressively reorient the approach to waste management in a radical way. Industrialists are taking responsibility for all discharges generated by their activities, including the products they manufacture. Lanoie and Plouffe (2008) point out that by obliging all producers to take back their products for dismantling in order to recover their components and reintroduce them into the production chain, the problem of saturation was to be resolved. This anticipated regulation explains the first wave of "environmental design".

Tools are then developed to enable the looping of material flows. These tools are design for recovery, design for disassembly and design for recycling. The objective is to design products to facilitate the recovery of their components in order to reintegrate them into the production process. The main methods are: reuse, re-employ, recycle and energy recovery. These tools lead to several recommendations to reduce the volume of waste:

– designing products using as little material as possible;

– increasing the life span of products;

– designing removable products;

– promoting the recovery of products at the end of their life;

– promoting the use of recycled materials in manufacturing.

Companies have therefore proposed design for recovery tools in the form of documents indicating assembly methods and principles of material compatibility that promote rapid assembly and disassembly of products (Lanoie and Plouffe 2008). The design for recovery tool is recognized as one of the key competencies of the circular economy for the design of a product or service (Pasquale et al. 2021). Nevertheless, this tool only considers one criterion (solid waste). The increase in environmental impacts, due to the energy needed to recover, dismantle and recycle products, is not analyzed.

A second approach, known as the "waste approach", was developed in the 1970s. It consisted of evaluating the product with a view to reducing the waste generated during its development on the production site rather than studying the waste generated at the end of production. The objectives were to limit the production of waste, to reduce the toxicity of the waste produced and to promote the recovery and treatment of waste. This tool was primarily a tool for managing business operations and less a tool for product design. As with the design for recovery, the environmental impact was analyzed based on a single criterion. For Lanoie and Plouffe (2008), the scope of waste studies is therefore limited to production sites. The global impacts generated by the company in the long-term and the end-of-life of the products are not taken into account.

A third approach, known as the energy approach, emerged following the 1973 oil crisis, the threat of depletion of resources and the debates on the limits to growth that arose from the publication of the Meadows report at the Stockholm conference in 1972 (Abrassart and Aggeri 2002). The use of energy by companies was strongly questioned. Companies and states therefore focused on energy savings for strictly financial reasons. Studies appeared on products about the energy consumed during their development. Lanoie and Plouffe (2008, p. 8) quote Millet (1995): "All decisions related to the environment can be made by using the energy criterion as a starting point." Thus, the best product from an environmental perspective was the one that consumed the least energy throughout its life cycle. The tool developed to do this kind of analysis was the energy content (Pole éco-conception 2017). The goal was to minimize the total energy throughout the product's life cycle. Everything inherent in the development of the product (materials, shaping processes, transportation, waste, etc.) was therefore translated into energy consumption. For the design, the energy content made it possible to compare the potentially usable materials in a quick and simple way.

The energy approach, like the two previous approaches, was based on a single unit of measurement (energy) in the evaluation of products or systems. The analysis remained partial, because it could be considered, for example, that the use of a low-energy material was the best solution in the development of a product, while the natural resources necessary for its manufacture were almost exhausted.

2.1.2.3. *History of multi-criteria approaches*

The first multi-criteria approaches appeared in the United States in the 1970s. The principle is that the ecological *optimum* is obtained by minimizing five criteria (energy, materials, solid waste, water and air emissions) during the development of a product or service.

Initially, the approach was based on the integration of material balances, used by the chemical industry, with the energy balances developed following the oil crisis, to obtain a more complete picture of the resources needed to develop products. Later came the REPA (Resources and Environmental Profile Analysis), in which, in addition to taking into account the resources used, the discharges generated by the production system under study were also considered. In the mid-1970s, the urgency of waste reduction was the focus of attention, and it was not until the early 1980s that interest in these analyses resurfaced. In 1979, American industrialists created the SETAC (Society of Environmental Toxicology and Chemistry), whose objective was to develop and promote tools to evaluate the environmental impacts of a technique or an activity. During the 1980s, materials, energy, solid waste, water and air emissions were integrated into impact analyses to obtain an increasingly complete vision of the impacts generated by products or systems. All of these analyses, which were called "ecoprofile", "ecobalance", "cradle-to-grave analysis" or "life cycle analysis", have gradually been grouped together under the name "life cycle analysis" (Fava 2005; Lanoie and Plouffe 2008).

In Europe, the environmental impacts of energy use reinforced the urgency of developing energy analyses that integrated, in addition to the quantities of energy expended, the types of energy used. In the mid-1980s, the first results of studies on the environmental impacts of certain products were published, such as the ecological inventory of packaging by the Swiss Federal Office for the Environment. The method used to obtain the "ecological loads" of the products studied was based on four criteria: energy consumption, air emissions, water emissions and solid emissions (Blouet and Rivoire 1995). From the 1990s onwards, energy studies progressively took into account environmental impacts and were integrated into multi-criteria approaches.

In 1991, after its ninth workshop, SETAC published a report ("A Technical Framework for Life Cycle Analysis"), which defined the first method to be used to conduct an LCA. SETAC then identified the steps that constitute LCAs:

– life cycle inventory;

– life cycle impact analysis;

– life cycle improvement analysis.

From this method, SETAC undertook case studies, which allowed it to build databases and refine its method. LCA has become the most widely used tool (Mendoza et al. 2017). It is a methodology standardized by ISO 14040 and ISO 14044.

The "Green Paper on Integrated Product Policy", published by the European Commission in 2001, introduced the polluter-pays principle by taxing the most polluting products. LCA was then presented as a tool for achieving eco-design. The other tool resulting from the multi-criteria approach is the environmental audit. Like the LCA, it is standardized by the ISO (ISO 19011, 2002). It is a tool for managing the environmental impacts generated on production sites. This tool, like the waste study seen above, is specifically linked to operating sites.

2.1.2.4. *A vision dominated by engineering*

The engineer's approach can be defined as the search for the best use of available knowledge to produce or make possible what people want to have or want to do (Karlsson and Luttropp 2006). The approach is pragmatic and synthetic. Thus, "the engineer designs technical systems whose control he or she seeks to ensure" (Abrassart and Aggeri 2002, p. 3). The first definition of eco-design in this literature can be dated back to 1994 according to a report published for the European Commission (Tukkerr et al. 2000). Eco-design refers to a process of developing environmentally friendly products. Van Weenen adds in 1995 that the design of new products or product improvements is achieved with specific environmental criteria. Van Weenen brings the term eco-design closer to the term "design for environment" developed in the United States. It is "a practice in which environmental considerations are integrated into product and process engineering design procedures" (Van Weenen 1995, p. 96). The American approach seems to have inspired Glenn Johansson and his co-authors, who propose, in one of the first reports on the state of the art of eco-design approaches, this definition:

> Eco-design refers to actions taken and activities carried out as a result of integrating environmental performance criteria into a product development project (Johansson et al. 2001, p. 8).

The same year, in one of the first French books on eco-design, Vigneron and Patingre propose a similar definition:

> The eco-design of a product or service consists of designing it in such a way as to limit the consumption of natural resources and optimize its impacts on the environment and on human health throughout its life cycle, "from the cradle to the grave" (Vigneron and Patingre 2001, p. 47).

These authors take a technocentric approach to eco-design. The aim is to know the environmental consequences of the choices made in order to control their global effects. Similarly, Le Pochat (2004) places himself in this perspective, eco-design is defined by its object: the minimization of the environmental impact. This objective appears clearly in the most commonly used definitions of eco-design in companies, given by the French standard NF and the ISO standard:

> Systematic integration of environmental aspects from the design and development of products (goods and services, systems) with the objective of reducing negative environmental impacts throughout their life cycle for equivalent or better service. This approach from the beginning of a design process aims to find the best balance between environmental, social, technical and economic requirements in the design and development of products (Standard NF X 30-264, environmental management: help in setting up an eco-design approach, 2013)[1].

> A systematic approach, which considers environmental aspects in design and development with the aim to reduce adverse environmental impacts throughout the life cycle of a product (ISO 14006 v2020 standard)[2].

Other authors in the field of engineering emphasize the technological objectives of eco-design. Thus, for Lifset and Graedel (2002), eco-design is a technological innovation at the micro-economic level. It testifies to an industrial optimism and a strong responsibility of engineers. Implicitly, eco-design is then in a life cycle perspective and relies on the cradle-to-cradle approach developed by Braungart et al. Roux et al. (2005) take a similar approach:

> A good eco-design will also aim to reintroduce into the consumption cycle the materials and energy that can be recovered at the end of their life cycle to close the loop (Roux et al. 2005, p. 3).

1 See: https://www.eco-conception.fr/static/definition-de-leco-conception.html.

2 See: https://www.eco-conception.fr/static/definition-de-leco-conception.html.

This involves developing products from a circularity perspective where products can then be recovered and reused consistently (Le Moigne 2014). Roux et al. (2005) distinguish between eco-evaluation and eco-technologies. Eco-evaluation consists of defining and implementing evaluation criteria. It is thus a step in technocentric eco-design. As for environmental technologies, they refer to technologies that generate less impact on the environment, use resources in a more sustainable way and allow for the recycling of all or part of their waste in a more acceptable way than the technologies they replace.

In addition to the design of a product with the objective of reducing its environmental impact throughout its life cycle, design for X approaches were developed in the 2000s with a defined objective such as recyclability, remanufacturing, reuse or reassembly (Arnette et al. 2014). This method is relevant when the company is looking for specific solutions. Samuel Mayer explains that she then identified what she wanted to work on (Box 2.2). "Design for" is then a tool that directs the results very precisely. Thus, design for remanufacturing consists of designing a product that can be disassembled at the end of its life (Arnette et al. 2014).

2.1.3. *Sustainable design*

Sustainable design is about combining environmental and social dimensions in a long-term perspective.

2.1.3.1. *Issues and founding principles*

Madge (1997) distinguishes three stages in the history of design and the environment: green design in the 1980s, with the greening of advertising campaigns and green marketing; ecological design in the 1990s, with the creation of the association Ecological Design in 1988; sustainable design, with the publication of the Brundtland Report, defining sustainable development. It specifies sustainable design as "the design of materials and products […] that respect living species and the ecology of the planet" (Madge 1997, p. 48).

Sustainable design thus refers to "the ability to support a domain extended to the entire design-production system and its exchanges with environments" (Antonioli and Vicari 2015). It is achieved in the greatest harmony possible with the living species and ecology of the planet. Some designers, concerned about the growing environmental impact of economic activity and large-scale industrial development, are seeking ecological and ethical alternatives. They are considering radical changes in favor of an ecological transition:

> Sustainable design can and should play an important role in the transition to sustainable production and consumption as key elements of quality of life (Spangenberg et al. 2010, p. 1488).

Sustainable design insists on "the development of new products, new services allowing to live in total harmony with nature and allowing a more sustainable lifestyle" (Janin 2000, p. 32). The products as they are traditionally conceived are then challenged: these designers put forward a design offering more functions with the lowest possible cost. The useless and the superfluous are banned. Janin (2000) quotes Manzini (1991a): "[Sustainable design] is then a design activity whose objective is to link what is technically possible with what is ecologically necessary in order to propose socially acceptable and culturally acceptable solutions."

Sustainable design is about innovation, using ecological principles, and encompasses ethical and social aspects (Knight and Jenkins 2009). For Arnette et al. (2014), design sustainability combines the three pillars of sustainable development. According to the authors, it enables social aspects such as the capabilities of people with disabilities and the elimination of child labor to be further integrated into the design process.

Finally, sustainable design can be associated with the "design of milieu" in the words of Petit (2015). "Milieu" refers to the design of a product or service for a given living environment. The milieu is considered here for its natural, technical and social dimensions. The object is conceived as an intermediary between the human species and its environment, so the object is thus in the milieu. Petit (2015) quotes Manzini (1991b), for whom in a perspective of environmental and social sustainability, sustainable design can lead to changes in the behavior of the design, production and consumption system. For example, Kazazian calls for a shift from a consumer society to a user society (Kazazian 2003). The development of the Autolib' (French electric car rental system) is an example of sustainable design combining product innovation and new uses, going beyond the technical aspects of the new electric motor (Petit 2015). We can also mention the activities of the New Alchemy Institute research center. In the mid-1970s, it created bio-shelters that allowed plants to be grown, fish to be raised, and a laboratory to be heated and lit by the sun from a design deliberately modeled on nature (Todd 2005). The residents of the New Alchemy Institute are changing their behavior: new ways of growing food, new ways of working in the search for greater harmony with nature. New Alchemy is consistent with Kazazian's approach:

> Nature can be a source of inspiration for new business strategies, capable of reinventing the relationship between man and nature and the object (Kazazian 2003, p. 6).

At the end of this analysis, our definition of sustainable design will be the following: the design of new products, new services for the human species in the respect of its living environment in a long-term perspective as a key element of its quality of life and the preservation of the planet.

2.1.3.2. *Creativity in the service of ecology*

Sustainable design is demanding, because it relies on the creativity of designers to go beyond the reduction of environmental impacts. We can then include the approach of Abrassart and Aggeri (2002) when they define eco-design as a process combining "new expertise and new environmental performance criteria in a design process" and the search for "compromises that meet the different requirements of the product specifications" and this "through cross-learning" (Abrassart and Aggeri 2002, pp. 15–16). This design approach questions "social uses, the economic and environmental value of product or technological concepts and the scientific and technical knowledge useful to incorporate" (Abrassart and Aggeri 2008, p. 250). The authors' analysis is consistent with our definition of sustainable design. Product design can follow two paths: on the one hand, with an artistic approach, according to which the form and the symbol are determining; on the other hand, with a "functionalist" approach, according to which the uses and technical concepts are essential. The interactions between the user and the product can be modified with dematerialization, reuse and the transition from product to service. According to Abrassart and Aggeri, these approaches generate "new interactions with potential customers to experiment and invent new uses" (2008, p. 264). We will see that the functional economy leads to forms of innovation that combine environmental benefits and new practices among users. For his part, Debref (2017) emphasized the importance of a creative approach in his definition of eco-design. This is, in fact, "a source of inventions and preventive actions". It then makes it possible to integrate directly new objectives and missions to prevent climate change (Debref 2017, p. 17). Similarly, Plouffe et al. (2011) highlight the contributions of a creative approach to respond to the challenges of sustainable development.

Figure 2.1, based on Spangenberger et al. (2010), compares technocentric eco-design and sustainable design. The objectives are different: the first one consists rather of improving what already exists with a prevention strategy. The second one aims to transform what already exists with a precautionary approach: it is to meet the needs of current and future generations in a long-term perspective, characteristic of sustainable development. Here, sustainable development encompasses economic, environmental, social and institutional dimensions. Objects can be challenged; they must be transformed. The search for truly alternative design methods is encouraged.

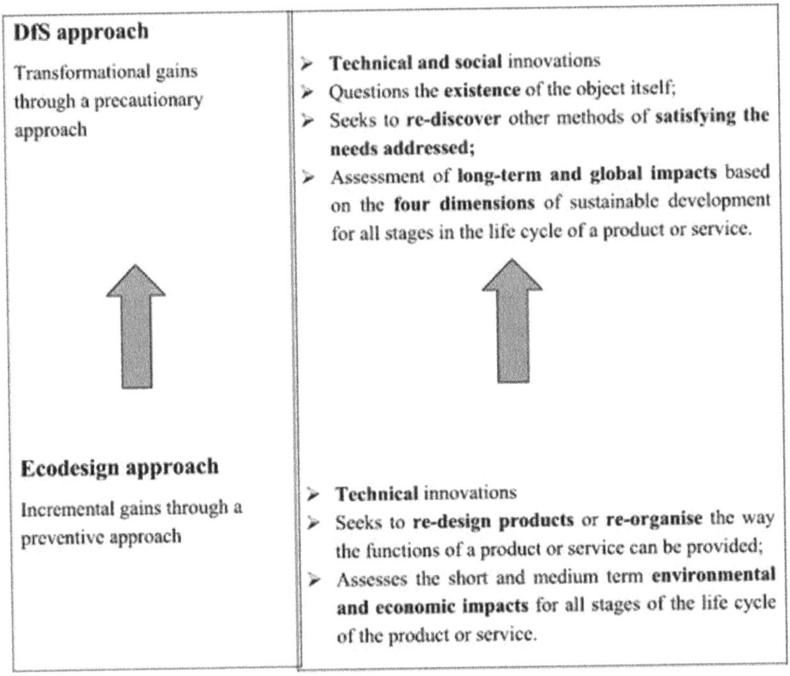

Figure 2.1. *Technocentric eco-design and sustainable design (from Spangenberg et al. (2010), p. 1488). For a color version of this figure, see www.iste.co.uk/vernier/ecodesign.zip*

2.2. Technocentric eco-design: a dominant approach

The technocentric vision remains dominant in eco-design methods (Janin 2000; Bellini and Janin 2019) and also in the literature in economics, management and in the practices of companies. It will become the reference in international ISO standards and for the European Commission. In a technocentric vision of eco-design, the ecological transition consists of finding techniques to maintain our way of life. The electric vehicle is an interesting example showing how eco-design can reduce environmental impact by considering technique as a way to adapt to the environment (Petit 2015). The consequences of eco-design are mostly limited to a change in the engine. In contrast, there is little impact on user behavior when switching from a combustion vehicle to an electric vehicle.

In the case of industry, several eco-design methods have been identified, with LCA being the most widely used tool in industry.

	Description of the tool
Multi-criteria life cycle assessment	Consideration of all stages of production from the use of raw materials to the end of the product's life, corresponding to an analysis of inputs and outputs within the production chain.
Single-criteria life cycle assessment	For example, product carbon footprint, water footprint.
Matrix approach	Analysis of the impacts according to a two-dimensional table: the first indicating the stages of the life cycle and the second the environmental impacts. For example, MET matrix.
Analysis of environmental effects	Technique from the quality management system. It works as the analysis of the failure modes and effects.
Qualitative approach	For example, simplified and qualitative life cycle assessment, which consists of improving the existing product. Based on the principle of benchmarking.
Integrated approach (EIME) or environmental impact assessment	Implementation of a continuous improvement strategy. It is based on the consideration of environmental impacts, the reduction of environmental risks, the preservation of health, natural resources, air and water and the limitation of waste.
Guidelines	In-depth analysis of impacts through a selection of strict steps within the product development process or life cycle.
List of substances	For example, REACH regulation, which indicates substances subject to authorization and restriction.
Checklists	List of questions to guide eco-design strategies. For example, eco-design strategy wheel, TRIZ method.
Standards	General approach (e.g. ISO 14006 standard, which gives guidelines for incorporating eco-design) and sectoral approach (e.g. NF ISO 22628 (2002) standard) in the automotive sector.
Other	Ecological backpack (seeking equality between inputs and outputs with an LCA) and financial approach (perform life cycle cost analysis, total costs, externalities and life cycle management).

Table 2.1. *Eco-design tools (from Debref (2018) and Bellini and Janin (2019))*

2.2.1. *The diversity of available methods*

Based on the work of Debref (2018) and Bellini and Janin (2019), 11 methods can be identified to assess environmental impact in an eco-design approach. They are presented in Table 2.1. This diversity means that designers can develop different assumptions and several outcomes for an initial product design situation. While this gives more flexibility in design choices, it increases the workload for designers to choose the most suitable tool, which can be a hindrance to its effective implementation (Knight and Jenkins 2009; Debref 2018). Finally, LCA is apparent in nearly one out of two tools.

2.2.2. *The most commonly used method: life cycle analysis*

The LCA is a tool for analyzing the environmental impacts of a product. The results obtained determine the nature of the modifications to be made in all phases of the product's life cycle to reduce its environmental impact. It is often mentioned as a relevant tool for identifying the environmental impacts associated with a product (Blouet and Rivoire 1995; Karlsson and Luttropp 2006; Plouffe et al. 2011; Aurez and Georgeault 2016a; Mendoza et al. 2017). The European Commission's Green Paper on Product Policy (Commission européenne 2001a) presents LCA as the tool to achieve eco-design. It is also the only tool that has been the subject of international standardization work (ISO 14040 and ISO 14044).

The analysis of the product's impact on the environment concerns the different phases of its life cycle: it implies simultaneously considering the impacts during the selection of raw materials, the industrial process, the storage and transport, the use and end-of-life phases. LCA is defined as follows by the ISO standard 14044:

> LCA addresses the environmental aspects and potential environmental impacts [...] throughout a product's life cycle.

For Brodhag (2014), eco-design is both a multi-step, multi-criteria and multi-actor approach:

– all stages of the life cycle are considered;

– the objective is to control all environmental impacts: air, water and soil pollution, greenhouse gas emissions and ozone-depleting substances, consumption of resources and energy;

– all the actors involved throughout the product's life cycle are concerned by this analysis.

The LCA methodology has four phases (Pôle éco-conception 2007; Aurez and Georgeault 2016a):

1) Definition phase of the object of study, the objectives pursued and the boundaries of the system that will be analyzed.

2) Inventory phase of the data to be evaluated. It is therefore necessary to establish the list of components and processes, as well as the quantities and nature of the materials, the energy used and the discharges generated.

3) Evaluation phase: this involves associating environmental impacts with components and processes and compiling them for comparison.

4) Interpretation of the results obtained by making recommendations to remedy the weaknesses identified.

During phase 1 (the definition phase), it is a matter of specifying the function performed by a product or service (Aurez and Georgeault 2016a). It also involves taking into account all the activities that take place during the five stages of its life cycle, from its design to its end-of-life (Box 2.1). The product's life cycle allows us to define the boundaries of the system under study. Each stage is broken down into elementary processes.

During phase 2 (inventory phase), a detailed accounting of the flows passing through the system is necessary. Initially, an inventory of flows is carried out. Second, the quantification and the collection of data on the impact of each component since its extraction phase must be carried out. Databases concerning common raw materials, energy, transportation, processes are therefore built to facilitate the designer's work (Aurez and Georgeault 2016a).

In phase 3 (evaluation phase), the designer links the impacts to the materials and energy used and the waste generated. The product or service will then be analyzed according to the five stages of its life cycle to ensure that all activities are taken into account. The stages are presented in Box 2.1. Each stage of the life cycle involves elementary processes (processes, activities, etc.) that require inputs (materials and energy) and generate outputs (solid, liquid, gaseous). The designer analyzes the inputs and outputs. For example, on the one hand, inputs can lead to the depletion of resources, and on the other hand, outputs can generate emissions that are harmful to the health of fauna and flora. The overall impact of the product or service is therefore attributable to both inputs and outputs. The analyst must quantify them and measure their impacts in order to identify the product's ecological weaknesses.

According to the life cycle assessment method, we must distinguish between:

– the incoming flows of materials and energy necessary for the realization of the various stages of the life cycle;

– outgoing flows, whether co-products, by-products, atmospheric emissions, liquid effluents, solid wastes and others.

All these flows are then translated into environmental impacts. ADEME currently distinguishes between the impacts listed in Table 2.2.

First stage (raw materials): this stage consists of the extraction of raw materials from their natural environment, their storage and transport to the place of transformation, the operations of their transformation into materials, their storage and transport to the place of manufacture (e.g. extraction of crude oil until the obtention of polypropylene beads ready to be used in a plastic injection process).

Second stage (manufacturing): this stage includes the storage of materials, their processing into a finished product, their packaging and storage and their transportation to the distributor's storage site. A product may consist of several components manufactured in separate locations. In this case, all modes of transportation and intermediate storage locations, as well as all distances traveled and the various types of packaging used, should be considered. In addition, this step should include component assembly operations (e.g. manufacturing and assembly of a polycarbonate visor on a protective helmet).

Third stage (distribution): this stage includes the activities of storage of the finished product and all the distances involved in transporting it to the point of sale, display/storage in the store and transportation to the point of use. This step must take into account the different means of transportation involved such as truck, train, ship or plane, as well as the energy sources used such as diesel or kerosene (e.g. handling retail wrapped chairs sent by train).

Fourth stage (use): this is the life cycle stage where the product performs its function, that is, the period during which its user benefits from the product. This includes the maintenance and repair of the product, as well as the supply of consumables; a consumable is a secondary product that is essential to the operation of a product (e.g. the use of a coffee maker requires the use of paper filters, water and energy). The behaviors and habits of consumers are key factors, as they can significantly influence the quantity and quality of consumables used. Therefore, these consumables and their impacts must be considered throughout their own life cycle.

Fifth stage end-of-life this stage considers the different end-of-life scenarios when the product or service is no longer able to fulfill its original design function or is no longer suitable for the user. This may include recovery (reuse, recycling or incineration) and/or landfill of the product at the end of its useful life and also of packaging and consumables.

Box 2.1. *The five stages of the product's life cycle (from Pôle éco-conception (2007))*

	Impact criteria
Impact indicator for air	– Contribution to the greenhouse effect. – Air acidification. – Tropospheric ozone formation. – Depletion of the ozone layer. – Particulates and respiratory effects of inorganic substances.
Impact indicator for water	– Eutrophication of fresh waters. – Aquatic ecotoxicity. – Eutrophication of marine waters. – Water consumption (flow indicator).
Impact indicator on soil resources and human health	– Primary energy consumption (flow indicator). – Depletion of non-renewable resources. – Depletion of limited energies and their sources. – Human toxicity. – Land use.

Table 2.2. *Impact criteria (from ADEME 2022a; Pôle éco-conception (2022b))*

Finally, phase 4 (the results interpretation phase) is associated with the other three phases. It involves the ongoing validation "that the results allow the study objectives to be met" (Aurez and Georgeault 2016a, p. 247). The scope of the study may be modified given the availability or unavailability of certain data.

A study report concludes the LCA and presents "the objectives and scope of the study, the limits and assumptions, the technical, temporal and geographical representativeness of the data used, the results of the study and their critical analysis" (Aurez and Georgeault 2016a, p. 247).

LCA makes it possible to identify impacts from a global perspective. The aim is to avoid a situation where the final product has a reduced impact on the environment at one stage of the life cycle but an increased impact at another stage. Margni et al. (2016) give the example of electric and thermal cars. Environmental emissions are displaced from the fourth stage (use) to the second stage (manufacturing) due to emissions associated with the power plant, distribution network and battery production.

The results of the LCA are analyzed and interpreted collectively by the eco-design project team (Debref 2018). These results allow the company to develop its product strategy (Box 2.2). Indeed, in theory, companies should analyze all stages of the product's life cycle to identify environmental impacts. In practice, companies select their environmental strategy based on the results of the LCA on the product and the consequences of reducing the environmental impact on each stage (e.g. costs, materials to be used, energy required, weight of the final product) (Berneman et al. 2009; Vernier et al. 2022).

In Box 2.2, we repeat the interview conducted on May 19, 2022 with Samuel Mayer, Director of the Pôle éco-conception. This sheds new light on the place of eco-design in the company's strategy, on the importance of the team around the eco-design project and on the close relationship between the circular economy and eco-design. These points are developed in this and subsequent chapters.

Could you briefly present the activities of the Pôle éco-conception?

The cluster was created in 2008 to support the development of sustainable consumption and production methods. As an association of industrialists, the cluster supports organizations to increase their performance and create value through life cycle thinking while reducing environmental impacts. It has four missions: to disseminate information on life cycle thinking, to support a network of intermediary structures and companies, to advise and train companies and to consolidate eco-design approaches by developing tools and methods.

What is eco-design?

It is a method that consists of taking into account the environment from the design stage. It relies on tools to evaluate the environmental impact, which can be an LCA. For a long time, LCA was considered the most complete tool. We are coming back to this, because LCA does not measure everything. How can we measure the impacts on biodiversity, for example? When we need a complete analysis, we can add elements to LCA.

What are the main steps?

First of all, let us specify that eco-design is part of the company's strategy. We have the following stages:

– the positioning of the project in the company;

– evaluation with an initial analysis to be done with different tools such as LCA;

– the search for solutions, that is, the search for possible scenarios given the identified impacts on materials and components;

– the choice between these scenarios and the identification of the environmental strategy produced;

– the realization of the comparative environmental analysis compared to the initial situation to control the impact transfers from one stage of the life cycle to another or from one impact indicator to another;

– communication on the product strategy.

The evaluation can, of course, be done with LCA. It is a balance sheet of incoming and outgoing flows. In order to understand these flows, we will formalize them in impact. In general, there are several indicators of impact. But evaluating these impacts is not carrying out eco-design. What counts is what the company does next. Marketing, strategic positioning, competitive positioning, legal elements, etc. must all be considered. Once the environmental aspects have been identified by the evaluation, whatever the tool, they must be selected, because the company cannot work on everything. Then, we move onto the search for solutions and then to the choice of scenarios. The company must identify the environmental aspects for which it has a level of control or influence. Then, it will choose the environmental aspects that are important because of a customer request or a legal requirement. The product environmental strategy is carried out by crossing the environmental assessment with the strategic positioning. In other words, the company will select the significant environmental issues and back them up with actions based on criteria (CO_2, rate of recycled material, removal of such a substance that disrupts the end-of-life substitution of components that appear to be endocrine disruptors, etc.).

Is it a real team effort?

Yes, indeed, to succeed in an eco-design approach, you have to manage a multi-functional team in the company. You have to identify the environmental issue or issues, and in particular the ones that can be controlled, as early as possible in the chain of actors. It will then be possible to observe high levels of improvement on the environmental criterion.

Various functions are involved in the product development process: marketing, strategic marketing, operational marketing, design, research and development and the design office for solutions. It is a job that involves different professions.

But doesn't the company always have the environmental skills in-house?

The company must choose an assessment tool that is adapted to the project and the company. In some cases, the company does not have in-house skills, so it requires the involvement of external expertise. Second, ADEME offers a free product assessment, which can be used by a motivated person with some knowledge of environmental issues to carry out a preliminary assessment.

What are the "design for X" strategies?

These are strategies for when the company is looking for solutions, once it has identified what it wants to work on. The company chooses the tool that directs very precisely, with a well-defined strategy of deployment. This is the design strategy for recovery or end-of-life.

Today, many companies are interested in increasing the life span. They must first make a calculation of reparability and therefore of durability. They have to identify the places where there can be breakages or failures, points of reliability and points of unreliability. Then, the company can set up a design to answer them and possibly a repair protocol. This can be done by doing it yourself with a specific partner or by returning the product.

According to ADEME, eco-design is a form of circular economy. What do you think about it?

I will answer in two points. First, the economy is not circular in all countries. When we talk about life cycle in the Netherlands, it is linear, it goes from raw materials to the end-of-life, and there is no closing of the loops. In France, the loops are closed. Second, the circular economy is based on loops to reduce waste. But the question of environmental performance is not asked. At the cluster, we are proposing to put in an indicator for monitoring performance. There must at least be an evaluation. The circular economy will look at economic performance in circularity loops. It aims to create loops and thus economic value. One idea would be to eco-design the circular economy to check that we are heading in the right direction.

Is the circular economy not enough to reduce the environmental impact?

Exactly, the entropy increases no matter what happens. Let us take a simplistic example: I am going to recycle expanded polystyrene and therefore run empty trucks on the road. It is likely that the amount of fuel expended will be greater than what can be recovered in terms of economic value and environmental value.

Moreover, recycling is limited while consumption continues to increase. On the one hand, there is the question of the resources available in the technosphere. Recycling is equal to the quantity put on the market minus the life span. And on the other hand, materials are recycled little or badly, so they generate additional pollution. For example, it is said that aluminum can be endlessly recycled, but energy must be added. It is always a question of entropy.

Do you have any alternatives?

The vision we have today is to continue to eco-design and do the same thing with less. But at some point, we are going to have to think differently about how to address service to people, service to the person, if we want to decouple consumption from sources of use. What can the company do? We think about production methods, but there are also consumption methods. An eco-designed T-shirt is fine, but if the customer only wears it seven times before putting it in the closet, what is the point?

We need to look at ways to make use available differently. Do we all need to have a drill in our house? On my street, there are 25 houses and therefore 25 drills, is this necessary? We need to think about whether we really want to decouple.

Also, you have to think about the composition of the product. Making a T-shirt from recycled cotton mixed with polyester is fine, but what to do with it at the end of its life?

Box 2.2. *Interview with Samuel Mayer, director of the Pôle éco-conception (May 19, 2022)*

2.2.3. *Some examples of technocentric eco-design*

Here are three examples to better understand the issues involved in eco-design. First, we present an example of life cycle analysis with the "Do Chair" chair by the American company Steelcase, and then two examples showing the changes in the design of a product and a service.

The life cycle assessment of the "Do Chair" chair by the American company Steelcase was carried out over the entire life cycle of the product, from obtaining raw materials to manufacturing, use and end-of-life. The analysis is divided into three stages:

– Upstream: it includes the components, the obtaining of raw materials and the associated manufacturing processes.

– Production: it includes the transportation of raw materials and components from suppliers to the Steelcase factory in Madrid, the product manufacturing processes and the treatment of waste.

– Downstream: it includes shipping to customers, maintenance, use of the product and end-of-life, both for the product and for the packaging.

The impact indicators are presented in Table 2.3. They relate to air (acidification potential, ground-level ozone formation potential), water (eutrophication potential, water scarcity potential), and land resources and human health (land use and transformation, abiotic depletion potential and abiotic depletion potential – fossil fuels), which are the criteria identified in Table 2.2.

The analysis is completed by indicators on:

– resource use: renewable and non-renewable primary resources, secondary materials, renewable and non-renewable secondary fuels and net freshwater use;

– waste production: hazardous, non-hazardous and radioactive waste;

– outgoing flows: components for reuse, materials for recycling, materials for energy recovery and energy export;

– other indicators: human toxicity, aquatic ecotoxicity and land use.

Eco-design most often concerns products. For example, we have the example of the company EPTA France, which manufactures display cases and refrigeration equipment for supermarkets (Pôle éco-conception 2018, fiche éco-initiative 12). The eco-design approach has generated changes at different stages of the life cycle: the choice of raw materials (choice of composite materials instead of metal structures), manufacturing (lower water consumption and reduction of assembly time), logistics (reduction of transport and storage costs), use (lower electricity consumption) and end-of-life (easier dismantling and sorting of parts). The eco-design approach has finally made it possible to strengthen the thermal insulation, reduce energy consumption, improve ergonomics for the user and increase the glass surface.

Parameter		Unit	Upstream	Production	Downstream	Total
Potential global warming	**Fossil**	kg CO_2 eq.	7.17E+01	5.93E+00	4.05E+00	8.17E+01
	Biogenics	kg CO_2 eq.	4.20E-01	9.27E-04	3.57E-04	4.21E-01
	Use and processing of land	kg CO_2 eq.	5.89E-02	1.06E-03	8.25E-05	6.01E-02
Total		kg CO_2 eq.	7.22E+01	5.93E+00	4.05E+00	8.22E+01
Acidification potential (AP)		kg SO_2 eq.	2.78E-01	1.76E-02	1.69E-02	3.13E-01
Eutrophication potential (EP)		kg PO_4^{3-} eq.	1.16E-01	3.13E-03	2.77E-03	1.22E-01
Tropospheric ozone creation potential (TOCP)		kg NMVOC eq.	2.30E-01	1.62E-02	2.34E-02	2.69E-01
Abiotic depletion potential		kg Sb eq.	3.89E-04	5.62E-07	3.51E-07	3.90E-04
Depletion potential abiotic – fossil fuels		MJ, net calorific value	1.13E+03	8.74E+01	5.74E+01	1.28E+03
Potential for water shortage		m^3 eq.	4.16E+01	8.38E-01	2.04E+00	4.44E+01

Table 2.3. *Life cycle assessment for the "Do Chair" (from Steelcase (2022))*

There are fewer cases of eco-designed services. We have the example of Groupe SEB, a French manufacturer of household appliances, which has developed an offer to rent cooking appliances to individuals (Pôle éco-conception 2016, fiche éco-initiative 10). The eco-design approach has led to changes in two stages of the life cycle: manufacturing (choice of reusable and more resistant packaging and increase in the number of spare parts) and logistics (creation of collection points in urban areas).

2.3. The contribution of eco-design to environmental innovations

In addition to eco-design, there are a variety of other practices used by companies to reduce the environmental impact of their activities. These include the innovations enabled by the circular economy and the functional economy, as well as the frugal innovations applied most often in developing countries.

In this section, we wish to define the nature of the benefits of eco-design. We will start with an analysis of the concepts of innovation and environmental innovation. Then, we will study how eco-design can be associated with innovations that can lead to goods or services with a reduced impact on the environment, by distinguishing between innovations enabled by the circular economy and the functional economy and frugal innovations. We will thus be able to identify which forms of eco-design (technocentric eco-design and sustainable design) can be associated with them.

2.3.1. *Innovation, environmental innovation and sustainable innovation*

Innovation is the introduction of a new or significantly improved product, process, marketing or organizational mode compared to those already on the market. The *Oslo Manual* defines four types of innovation: product innovations, process innovations, marketing innovations and organizational innovations (Box 2.3). An eco-design approach can lead to these four categories of innovation. However, in the literature, product innovation is most often cited.

Product innovation: the introduction of a new good or service. This challenge includes significant improvements in technical specifications, components and materials, embedded software, user-friendliness or other functional characteristics.

Process innovation: the implementation of a new or significantly improved method of production or distribution. This notion implies significant changes in techniques, equipment and/or software.

Marketing innovation: the implementation of a new method of commercialization involving significant changes in the design or packaging, placement, promotion or pricing of a product.

Organizational innovation: the implementation of a new organizational method in practices, workplace organization or the firm's external relations.

Box 2.3. *The four types of innovation (from OECD and Eurostat 2005)*

Schumpeter defines innovation as a new productive combination that appears continuously or discontinuously (Schumpeter 1981). In other words, innovation can be either incremental, in which case it consists of a continuous improvement of existing technological systems (improvement of an existing product, replacement or modification of components) or radical, in which case it is based on a discontinuous improvement and breaks with existing technological systems (Rennings 2000). In Chapter 4, we will question the capacity of eco-design to lead to incremental and radical innovations.

An innovation is termed environmental when it reduces environmental impacts (Boutillier et al. 2012). In general, environmental innovations are most often considered in a technocentric approach: humans adapt to the environment through technology (Petit 2015). The social aspects are little considered. We will distinguish two typologies that allow us to classify environmental innovations. A first typology is that of Rennings (2000). He identifies two types of definitions to characterize the environmental dimension. It concerns either the meaning of the innovation or its content. In the first case, environmental innovation aims to "reduce or manage negative environmental impacts and/or maintain or improve positive environmental impacts" (Rennings 2000; Gasmi and Grolleau 2003). It is typically a product resulting from a technocentric eco-design approach with the objective of reducing environmental impact. In the second case, environmental innovation refers to new or improved processes, equipment, products, techniques or management systems that avoid or reduce environmental impact (Rennings 2000; Arundel et al. 2007; Horbach 2008). For instance, these include processes for cleaning wastewater.

A second typology is that of the OECD. It allows for the analysis of innovations leading to social change and sustainable design approaches. The OECD typology is built on three dimensions (OECD 2009). The first dimension is based on the traditional categories of innovation according to their objectives: product innovations, process innovations, marketing innovations (or marketing methods) and organizational innovations, to which the OECD adds institutional innovations (socio-cultural norms and institutional structures). The second dimension includes four categories of innovation mechanisms: modification, redesign, alternative innovation and creation. The third dimension incorporates the extent of environmental change. Figure 2.2 shows that the environmental innovations with the greatest impact are at the top right: these innovations are institutional innovations, based on new products (creation) and with a high environmental impact. These innovations require actors to coordinate. Radical innovations, with the highest environmental benefits, constitute the last phase of innovations.

The OECD typology allows us to have a broader view of environmental innovations. Innovations with organizational and institutional objectives and based on creation mechanisms constitute the innovations with the highest environmental benefits according to the OECD typology. Organizational and institutional changes contribute to social changes. Environmental innovation is therefore similar to social innovation. This refers to "any innovative approach, practice, intervention, or product or service that has found acceptance at the level of institutions, organizations, or communities and whose implementation solves a problem, meets a need, or fulfills an aspiration" (Grison et al. 2018). It constitutes changes to improve human happiness (Hillier et al. 2004) or to promote the well-being of individuals and communities (Cloutier 2003). This is one aspect of the social dimension of sustainable development. This innovation can then be at the origin of social transformations. It represents an opportunity to define new social relationships or to redefine old ones. It allows the creation of links and interfaces between individuals or groups, organizations, institutions and territories (Fontan 2008). It also allows for new collaborations between individuals, structuring between individuals and structuring on a local scale. It thus responds to certain needs of the population outside the market and the field of intervention of the State (Hillier et al. 2004).

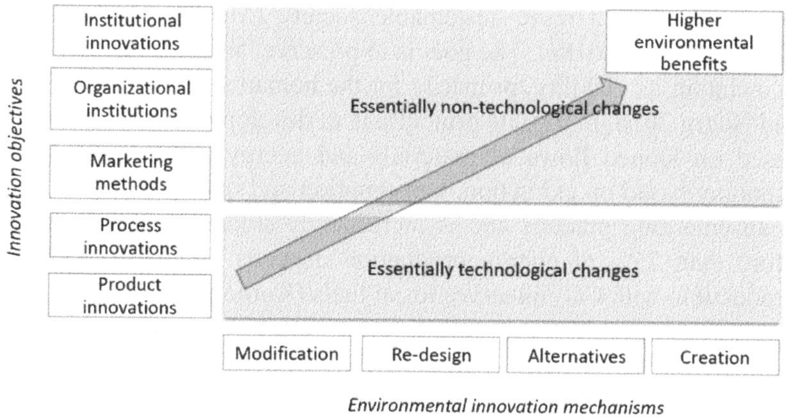

Figure 2.2. *Typology of innovation (from OECD 2009)*

Finally, these definitions of environmental innovation and social innovation can be combined in the notion of sustainable innovation. In the first chapter, we defined the social aspects of sustainability (the search for well-being, inter- and intra-generational equity, poverty reduction and the distribution of the gains from growth between activities). Sustainable innovation then integrates the needs and vision of the company, as well as the interests and needs of all parties involved in its activity (Le Bas 2017). Unlike environmental innovation, which is most often considered from a technical point of view, sustainable innovation takes into account the human dimensions of the company (Le Bas 2017) and also the users of the innovation (Trischler et al. 2020). The notion of sustainable innovation is close to the notion of sustainable design as we have seen previously.

2.3.2. *The circular economy and the functional economy: practices that are sources of environmental innovations*

2.3.2.1. *The circular economy: practices leading to environmental and, more often than not, technological innovations*

The circular economy is a multidisciplinary field that encompasses different approaches, methods and tools with the goal of reinforcing change towards a more sustainable society (Mendoza et al. 2017; Korhonen et al. 2018a). The goal is to preserve "the biosphere in order to maintain its viability, primarily for the human species" (Arnsperger and Bourg 2016, p. 97). Its principle is to develop economic activities based on looped flows of materials and energy instead of a linear economy based on extraction, consumption and release as waste. This is an emerging practice and is increasingly studied in the literature. More than 75% of energy production depends on linear modes of production, with CO_2 intensive fossil fuels (Korhonen et al. 2018a).

The circular economy has many definitions. They underline systemic issues (Aurez and Georgeault 2016a; Geissdoerfer et al. 2017; Mendoza et al. 2017; ADEME 2022b), the reduction of environmental impacts (Geissdoerfer et al. 2017; ADEME 2022b) and the search for efficiency (Mendoza et al. 2017; Lüdecke-Freund et al. 2019; ADEME 2022b). Some authors take a product life cycle perspective (Aurez and Georgeault 2016a; ADEME 2022b). We provide some definitions in Table 2.4 showing the different approaches.

The development of the circular economy leads to new economic activities. For Kirchherr et al. (2018), these activities replace the principles of end-of-life of a product with those of reducing, reusing and recycling of materials in the processes of production, distribution and consumption processes. For Korhonen et al. (2018b), the circular economy takes the form of high-value material loops alongside traditional recycling activities and calls for developing sustainable consumption and production patterns.

Authors	Definitions of the circular economy
Aurez and Georgeault (2016a, p. 18)	The circular economy aims to "ensure the protection of biodiversity and a development conducive to the well-being of people". This requires "systematically reducing the amount of primary raw materials and energy at all stages of the life cycle of a product or service, at all levels of organization of a society".
Geissdoerfer et al. (2017, p. 759)	"A regenerative system in which resource inputs and waste, emission and energy leakage are minimised by slowing, closing and narrowing material and energy loops."
Mendoza et al. (2017, p. 526)	"The circular economy aims at maximizing resource efficiency. The goal is to get out of the linear economy by preserving and enhancing natural capital, by optimizing resource yields through the circulation of products, components and materials at the highest utility within the loops and by fostering system effectiveness by revealing and designing out negative externalities."
ADEME (2022b)	"An economic system of exchange and production that, at all stages of the product (goods and services) life cycle, aims to increase the efficiency of resource use and decrease the impact on the environment while developing the well-being of individuals."

Table 2.4. *Definitions of the circular economy*

The precise description of circular economy practices differs among authors (Table 2.5). The functional economy or product-service systems can be integrated (Ghisetti and Montresor 2020; ADEME 2022b). We will study the functional economy in more detail in the following pages. Eco-design is also indicated. Thus, the category "design and reassemble sustainable products" (Ghisetti and Montresor 2020) is a form of eco-design oriented towards the sustainability of products. The category "redesigning products and services to minimize material use or the use of recycled materials" is a form of eco-design oriented towards waste and recycling (Garrido-Prada et al. 2021). ADEME also mentions eco-design.

Authors	Circular economy activities
Ghisetti and Montresor (2020)	– Prioritizing regenerative resources. – Converting and eliminating waste. – Designing durable and reassembled products. – Combining products and services with pay-per-use or more generally product-service systems.
ADEME (2022b)	Offer of the economic actors: – extraction/exploitation and sustainable procurement; – eco-design (products and processes); – industrial ecology and territorial ecology; – functional economy. Consumer demand and behavior: – responsible consumption (purchasing, collaborative consumption and use); – extension of the duration of use (reuse, repair and recycle). Waste management: recycling (material and organic).
Garrido-Prada et al. (2021)	– Replanning energy usage to minimize consumption. – Using renewable energy. – Minimizing waste by recycling or reusing waste or selling it to other companies. – Redesigning products and services to minimize the use of materials or usage of recycled materials. – Replanning the way water is used to minimize usage and maximize re-usage).

Table 2.5. *Circular economy activities*

We will rely on the analysis of Bocken et al. (2016), as they structure circular economy activities according to three types of material and energy flows:

1) Slowing down the material loops. The objective is to slow down the flow of resources by increasing the duration of use of the product or by using it more intensively. This can be done by designing more sustainable goods and extending the life of the product with service loops that extend the life through repair or reconditioning. Reuse and remanufacturing also make it possible to moderate the flow of

materials. Industrial companies are developing solutions to slow down material loops. For example, the French firm Remober[3] offers companies the opportunity to buy back their old electronic equipment in order to repair it and give it a second life.

2) Closing material loops: recycling closes the loop after the good is used. The principle of material recycling became widespread following the writings of Braungart et al. (2007), who developed the concept of cradle-to-cradle for organic and technological materials (Box 2.4).

> Braungart et al. (2007) developed a new philosophy based on the product's life cycle (Debref 2018). These are new models for product design and industrial processes in which material becomes a nutrient. The continuous circulation of material is made possible using two methods: biological metabolism and technological metabolism (Braungart et al. 2007).
>
> In biological metabolism, biological nutrients circulate. These are products that can be consumed during their lifetime. These nutrients for the living system return to the natural environment. In technological metabolism, materials can remain in a closed system of manufacture, recovery and reuse conserving their highest value through product life cycles.

Box 2.4. *The "cradle-to-cradle" approach*

3) Shrinking material flows: using less material for a product. This means using resources more efficiently. This is an element of the ADEME (2022). However, over time, this may lead to a more rapid use of resources rather than a saving of resources. In other words, the resources may no longer be available for future generations.

Industrial ecology and circular economy are close concepts (Gallaud and Laperche 2016). According to ADEME, the former is a component of the latter. For Bruel et al. (2019) and Lüdeke-Freund et al. (2019), industrial ecology is the founding concept. Both

3 See: https://remober.com/fr.

approaches are based on the principle of material circularity. Nevertheless, industrial ecology is only concerned with production processes. It is a systemic approach based on closed industrial ecosystems: the waste of certain industries becomes the inputs of other industries (Frosch and Gallopoulos 1989). Material loops thus develop between industries. Gallaud and Laperche (2016) explain that this is an ancient phenomenon. At the beginning of the 20th century, waste streams were circulating between industrialists. Companies such as Western Electric, DuPont and General Electric were members of the American Industrial Waste Trade Institution created in 1913.

Industrial ecology consists of material, energy and information flows that must function as biological ecosystems (Buclet 2011; Petit 2015; Gallaud and Laperche 2016). It aims to loop the system around itself as much as possible (Vivien 2003). Environmental technologies are then essential to reduce the resource constraint: they must optimize the consumption of materials and energy and transform waste into resources.

	Description
Reemployment	Identical use of the property by a new user.
Reuse	Use of the property for another purpose.
Remanufacturing	Reuse of end-of-life products and components as resources to manufacture new, identical or even more efficient products.
Recycling	Recovery operation whereby waste, including organic waste, is reprocessed into substances, materials or products for its original use or for other purposes.

Table 2.6. *Keywords for the circular economy (from Aurez and Georgeault (2016a) and Bocken et al. (2016))*

We can analyze this innovation from the OECD typology. The objectives of the circular economy concern products and processes. It consists of developing activities based on resource loops (materials and energy) in the form of reuse, remanufacturing and recycling, and which are no longer discarded as waste at the end of their life (Table 2.6). The mechanisms of these innovations are product modifications or redesigns. The circular economy thus leads to technological innovations aimed at reducing environmental impact.

The eco-design method is often associated with the circular economy. The nature of the relationship varies according to the authors. It can be a prerequisite step for the development of the circular economy (O'Brien et al. 2014; Vence and Pereira 2019). We have seen that it is also a component of the circular economy by extending the life span of products, slowing down material loops (Commission européenne 2016; Ghisetti and Montresor 2020; Garrido-Prada et al. 2021). For ADEME, it is one of the seven pillars of the circular economy. Finally, according to De Perthuis (2014, p. 25), it is one of the levers of the circular economy, which "consists of integrating, from the design of products, the greatest possible number of parameters concerning the optimization of their uses throughout their life cycle: manufacture, use, disposal". We will return to the differences between these approaches in Chapter 4. We can already emphasize that the objective of eco-design is the reduction of environmental impacts, whereas that of the circular economy is the elimination of material and energy waste.

2.3.2.2. *The functional economy: how to combine environment, technology and new user behavior?*

A concept that dates back to the late 1980s, the functional economy aims to optimize the use of goods and services by maximizing their use value and minimizing the consumption of material resources and energy. The company no longer sells a good but its use (Gallaud and Laperche 2016). The functional economy thus leads to proposing a new business model for the company (Vence and Pereira 2019). It can be associated with the "product-service system" (Laperche and Merlin-Brogniart 2020) with joint marketing of products and services in a variable proportion given the "needs of the user without transferring the ownership of the good but providing its functionality for a given period of time" (Vence and Pereira 2019, p. 2).

The user is thus decisive in the design of the service offered. We can cite the case of the company Les Etablissements Cros, based in Échirolles (Isère), which historically sold equipment in the fields of industry and construction. Under the pressure of demand from a few customers, it modified its offer concerning compressed air. Instead of selling machines, it only sells cubic meters of compressed air. Thus, on the one hand, it is the needs of the users which explain this new business model, and on the other hand, it makes it possible to prolong the life of the equipment, because the company can more easily carry out maintenance or repair operations. According to ADEME[4] data, the company has reduced its carbon footprint compared to a conventional offer, with improved maintenance, less travel for technicians and longer machine life. A second example that illustrates the decisive role played by users is the rental of products. For example, in 2015, the company SEB launched the Eurecook service, which allows individuals to rent cooking appliances. SEB, which owns the appliances, can maximize the use value by increasing the number of users for the same product and extend the life of the asset. Once returned by the consumer, the products are cleaned, verified and repackaged. Today, this service is being expanded with the RepareSeb[5] project.

Finally, the eco-design approach on a product can contribute to the functional economy. Indeed, the designer, by seeking a reduction of environmental impacts, can opt for the functional economy. Innovation is based on creativity, which enables new practices to be found for users (Abrassart and Aggeri 2002). In other words, for the same amount of material resources linked to the production stage, usage will be increased. Moreover, the firm that retains ownership of the good can increase its life span by changing worn parts more easily, repairing and checking the good's proper use (Aurez and Georgeault 2016a). This strategy thus relies on the manufacturer's capacity to maintain, repair and recycle the asset over its lifetime (Vaileanu-Paun and Boutillier 2012b).

Based on the OECD typology, the objectives of the functional economy can be described as organizational and institutional. The mechanisms concern alternatives or the creation of new services. The

---

4 See: https://bo.longuevieauxobjets.gouv.fr/sites/ademe.fr/files/documents/FICHE%20ANDRE%20CROS.pdf [Accessed September 14, 2022].

5 See: https://www.groupeseb.com/fr/repareseb [Accessed September 18, 2022].

functional economy is thus a source of technological and social innovation due to new practices among users.

2.3.3. *Frugal innovation: a form of sustainable design*

Frugal innovation is a process of innovation and design in which user needs are key to proposing appropriate, adaptable and accessible products and services, especially financially (Le Bas 2017). Tiwari and Herstatt (2020) highlight three characteristics: be highly accessible financially, enable more efficient use of resources and offer high-quality solutions. The so-called cheapest car in the world, the Tata Nano, is an example of a frugal innovation offering a clever and financially accessible solution, especially because the technologies are inexpensive (Chia and Nayak 2012; Weyrauch and Herstatt 2017).

Frugal innovation thrives in markets with severe constraints on resources external to the firm: difficulty accessing inputs to produce, low consumer income and poor-quality infrastructure (Asakawa et al. 2019). While reducing environmental impact is only one of the goals of frugal innovation (Le Bas 2017; Hossain 2020), it nevertheless leads to limited environmental impacts. Given the difficult access to resources, it consists of using them better and thus saving them (Tiwari and Herstatt 2020; Le Bas and Khan 2021). It also leads to more robust, repairable products with a longer life span (Tiwari and Kalogerakis 2020; Levänen et al. 2022).

The design process integrates the characteristics of the production (availability of resources) and consumption systems (user's ability to buy, to appropriate the product or service). This approach has points in common with sustainable design, leading to innovations that combine environmental and social characteristics. On the one hand, the search for economies of resources, which consists of offering more functions with the lowest possible cost, makes it possible to avoid the unnecessary and the superfluous. We find the analysis of Antonioli and Vicari (2015) about sustainable design: materials and tools must be used as best as possible in order to "resolve a design problem at the lowest cost" (Antonioli and Vicari 2015, p. 101). On the other hand, the first frugal innovations appeared in low-income countries. Hossain

(2020), based on a study of three cases of environmental innovations in India, shows that they empower women, reduce the difficulty of manual labor and create business opportunities for marginalized people. Finally, frugal innovation can be brought closer to sustainable innovation (Le Bas 2017) for two reasons: on the one hand, it has environmental benefits, and on the other hand, it responds to the care of the most fragile individuals while allowing the company to achieve sufficient economic results.

If we take up the OECD typology, frugal innovation leads to product and process innovations. Innovation mechanisms are of the alternative or creative type, because they are based on products or processes that must be accessible and that everyone must be able to appropriate. Frugal innovation is a radical innovation based on emerging markets (Hossain 2020).

2.4. Conclusion

Under the term eco-design, we can identify two similar approaches. The first is technocentric eco-design, according to which innovation is enabled by technological progress and measured by the reduction of environmental impact. Its sustainability is weak. The second is sustainable design, a source of social and environmental division, and in which the search for harmony between the human species and its environment is a determining factor. It is part of a highly sustainable approach.

The technocentric vision of eco-design dominates companies' practices. Eco-design is often summarized as LCA, which enables the evaluation of environmental impacts. The company then has several design scenarios defined, for example, by the materials used and their impacts. It can then choose between these different scenarios. Finally, an innovation is qualified as environmentally friendly because its environmental impact is reduced relative to existing products on the market. It is thus most often incremental (Debref 2018). A technology is considered preferable relative to other existing technologies (Vivien et al. 2019).

On the contrary, sustainable design can lead to changes in behavior (Petit 2015), as shown by the example of Autolib': innovation is not only environmental, but also leads to new uses. It can contribute to the improvement of quality of life (Spangenberg et al. 2010). Finally, it can contribute to a change in design, production and consumption systems. In other words, eco-design can be the source of sustainable innovations that combine environmental and social benefits.

	Objective	**Main characteristics**
Innovations related to the circular economy	Remove waste during production cycles	Regenerative system. Eco-design can be a preliminary step. Technocentric approach.
Innovations related to the functional economy	Make available the functions associated with the products	The product remains the property of the company. User–company relationships are strengthened. Eco-design can be a preliminary step. Technological approach. New user behaviors.
Frugal innovations	Doing more with less	Design focused on essential functionality and accessibility. No eco-design. Technological and social approach.

Table 2.7. *Objectives and main characteristics of innovations related to the circular economy, the functional economy and frugal innovation*

Eco-design can be associated with different forms of environmental innovations. We present the objectives and the main characteristics of the innovations studied in this chapter in Table 2.7. We have been able to associate eco-design with the circular economy and functional economy practices. Innovations related to the circular economy can be associated with the technocentric vision of eco-design.

On the contrary, the new practices associated with the functional economy and the frugal innovations can be linked to sustainable design. If we take up the OECD typology, these two innovations are the most ambitious in terms of objectives and mechanisms for innovations.

In this chapter, we have defined eco-design and identified the link with the different forms of innovation. We will now study the practices of companies in terms of eco-design and the context in which they appear.

3

Eco-design in Companies: Practices and Decisive Actions

In 2010, in a survey of 400 companies, 20% of French companies declared that they were systematically integrating an eco-design process, 13% were starting the process, 29% were interested and 38% were not interested or did not feel concerned (ADEME 2022c). Eco-design practices were therefore a reality for a third of the sample. In the previous chapter, we studied how eco-design as a method of innovation had developed as a response by industry to environmental issues. Two categories of eco-design are present: a technocentric approach in which LCA dominates and a sustainable design combining social and environmental dimensions.

The objective of this chapter is to evaluate the contribution of eco-design approaches to the ecological transition. In a systemic approach, we will analyze eco-design practices in order to assess the extent to which there are factors favorable to the ecological transition and what type of eco-design is encouraged.

In the first section, we will present the systemic approach. In the second section, we will analyze the characteristics of the socio-technical systems in which eco-design approaches have been developed by distinguishing the elements of the socio-technical systems, the relationships between these elements and the coordination of the activities by the institutions. In the third section, we will study two visible aspects of the dynamics introduced by eco-design in the socio-technical systems: on the one hand, the economic results of the companies which can be significant of the acceptance by the market of these approaches, and on the other hand, the environmental communication which can promote the diffusion of these practices.

3.1. A systemic approach to innovation

Companies evolve within several systems: technical, sectoral, industry, national and territorial. They are in a permanent relationship with the different elements that make up these systems. This literature allows us to study the conditions in which innovations emerge and the strategies of actors in a system (Coppin 2002; Geels 2004; Carayannis and Campbell 2017).

The dictionary of the Académie française defines the notion of system as a set of diverse elements forming a whole in which all the parts are linked. In the systems approach, the observer is interested in "the organization that results from the structure of the relationships between processes and components and relates organization to the behavioral property of the whole" (Godard 2015, p. 61).

The innovation literature is based on the systems approach (Lundvall 2016). The authors use the concept of a national innovation system to explain growth. They define it as comprising "organizations and institutions engaged in research and exploitation, such as research and development departments, technological institutes and universities" (Lundvall 2016, Chapter 4, p. 97)[1]. More recently, for Laperche and Uzunidis (2007, p. 70), the national innovation system is defined as the "set of public and private institutions involved in the realization of innovation processes and related to each other by financial and information flows and by the movements of people". Individual, organizational and interorganizational learning is crucial in the system for understanding the link between innovation and growth. Thus, "not only the scientific infrastructures, but also the organizations that support the construction of competencies in labor

1 The book published by Bengt-Åke Lundvall in 2016 brings together 12 papers written over a period of 30 years (1985–2015).

markets, education, and professional activity" are major elements of the system (Lundvall 2016, p. 224).

The analysis of the concept of system allows the authors to highlight the importance of the relationships between the actors in the system and the system's own dynamics. Thus, for Carayannis and Campbell (2017, p. 176), the system is defined by "its elements and its own rationality: systems have a mode of operation, their own rationality, which organizes the self-organization and reproduction of the system and the relationships between the elements within the system and, moreover, the relationships between the system and other systems". For Coppin (2002, p. 34), the systemic approach makes it possible to analyze the "capacity of the actors in the environment to modify and adapt their behavior in response to changes in their environment".

In the innovation literature, the term system is used at different scales of analysis ranging from local to global (Carayannis and Campbell 2017). For example, according to Hekkert et al. (2007, p. 414), the innovation system refers to "all social subsystems, actors, and institutions that contribute in some way, directly or indirectly, intentionally or unintentionally to the emergence or production of innovation". Institutions and economic structures are studied to explain the technological change (Hekkert et al. 2007).

Finally, the systemic analysis of innovation integrates social innovation. Thus, for Lundvall (2016, p. 80), "institutional change and the strengthening of the skills and power of end-users" are social innovations, characteristic of the society in which they develop.

3.1.1. *Socio-technical systems*

The notion of socio-technical systems makes it possible to study the relations between the actors in a system formed around a technique and the way in which new techniques can emerge. This analysis is part of an evolutionary perspective and in the long-term (Geels and Schot 2007). It makes it possible to identify the conditions in which firms develop innovations. It thus constitutes a relevant framework for analyzing the conditions for the development of environmental innovations introduced by eco-design approaches. Even though this analysis is mainly technological, it makes it possible to integrate the strategies of organizations on the market and to analyze the conditions favorable to innovations.

The concept of socio-technical system refers to “the production activity of companies and the production, diffusion and use of technologies in relation to societal functions: transport, communication, food” (Coenen and López 2010, p. 1152). The reflection is conducted at three levels. The first level is micro-economic, where technological niches are found, places of incubation of radical innovations, innovation is protected from the market. The second level is that of the socio-technical regime, or technological regime in the sense of Nelson and Winter (1982), which designates the habits and practices shared by a community of engineers, and more broadly by scientists, politicians, users and interest groups. This highly structured set provides stable rules for the elements of the system (Coenen and López 2010). A socio-technical regime stabilizes the trajectory by encouraging actors to maintain the technologies in use rather than develop new ones driven by technological niches:

> First, technological niches are there to conserve and reinforce the longevity of environmental innovation in the face of dominant design (Debref 2018, p. 57).

Finally, at the higher level, the niches and actors of the socio-technical regimes are located in a socio-technical landscape. This is an exogenous environment for the actors in the system, which is difficult for them to change and which directs them towards certain actions (Geels and Schot 2007). The diffusion of niche innovations is possible under certain conditions according to Geels (2014): on the one hand, when these provoke internal dynamics due to learning processes, improvements in price and performance, as well as the support of certain groups of powerful organizations; on the other hand, when changes at the landscape level create pressures on the regimes that are opportunities for the diffusion of niche innovations. We then witness an alignment between the three levels favorable to the niches. This is how radical green innovations can emerge in conventional markets as they go against the dominant regime. Radical innovations are essential for companies:

> Niches are important because they provide places for learning processes about e.g. technical features, user preferences, public policies, and symbolic meanings (Geels 2004, p. 912).

The transition from a niche innovation to the technical regime is hindered because of the dependence on the technical pathway and the stability of the system around the dominant regime (Debref 2018). The market or industry is characterized by a dominant design with some form of supremacy (Haurand and Stummer 2018) and hinders the emergence of different products or processes. Dominant design indicates the key attributes of a product for manufacturers competing in a product class (Anderson and Tuschman 1990) and thus reinforces standardization (Abernaty and Utterbach 1978). In particular, the dominant design has features that respond to customer preferences. Future technical progress will then consist only of incremental innovations. Once a dominant design emerges, organizational and collective forces shape norms: these are the industry networks and market actors such as suppliers, distributors or customers, but these effects are limited (Anderson and Tuschman 1990). These norms do not allow niche innovations to emerge in the technological regime.

3.1.2. *Institutions*

Socio-technical systems analysis makes it possible to study how the elements of the system, and in particular the organizations, act within the framework of institutions that make it possible to coordinate and structure the activities (Geels 2004). The institutions can be likened to rules, since their purpose is the coordination and structuring of activities. They determine the relations between the elements of the system. They are constraints elaborated by individuals that structure political, economic and social interactions (North 1991). These institutions can be regulatory, normative or culturocognitive (Scott 2008).

Regulatory institutions are formal and informal pressures from system actors on other actors. They result from explicit regulatory processes. They are based on sanction and compliance. They include laws and technical standards (Geels 2004; Geels and Schot 2007; Coenen and López 2010).

Normative institutions are systems of values and norms that result in obligations, prescriptions and evaluation in social life. Value can be defined as what is preferred or desirable. Individuals compare and evaluate structures and behaviors according to their values. Norms define how to act and the legitimate means by which individuals can achieve their goals.

The cultural and cognitive institutions constitute the framework in which a meaning is given. The construction of social reality for each individual is based on culture and knowledge (Dequech 2003). Symbols such as words, concepts, myths, signs and gestures influence individuals by forming the meanings attributed to objects and activities (Geels 2004). In companies, cognitive structures influence the judgments and perceptions of individuals about the contexts in which they operate. Thus, engineers are influenced by cognitive routines and technological paradigms (Geels 2004). Cultural and cognitive institutions can be models, classifications and beliefs.

Rules or institutions in the sense of Scott (2008) contribute to the legitimacy of organizations in the system: a key dimension that reflects the acceptance of an entity in its environment (Suchman 1995), and in particular in its target audience. Organizations adopt the most socially accepted practices, rather than those that are most economically appropriate (Meyer and Rowan 1991).

3.2. Eco-design in socio-technical systems

Eco-design is a product design method that has led to environmental innovations in various socio-technical systems: we have given examples in the furniture (Steelcase), machinery and equipment (EPTA) and electro-mechanical (Groupe SEB) sectors. We question the conditions of the diffusion of eco-design in socio-technical systems. We will distinguish the way in which the actors of the systems participate, the relations between the actors and then the institutions that coordinate the systems.

3.2.1. *The elements of the system*

The production, use and distribution of the artifacts that are goods and services are the three spheres in which the fundamental elements of socio-technical systems evolve (Geels 2004; Figure 3.1). The elements of each sphere are:

– production sphere: organizations, production artifacts, scientific and technological knowledge, knowledge transfers from training centers, human resources, natural resources, machinery and tools;

– sphere of artifact use: repair and maintenance systems, cultural meanings, and artifacts complementary to a technology;

– sphere of distribution through markets, networks and infrastructures. This sphere connects the spheres of production and use of artifacts.

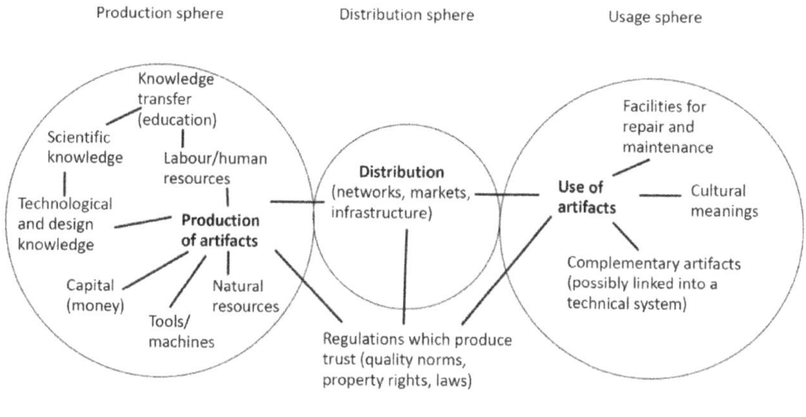

Figure 3.1. *The elements of the socio-technical system (from Geels (2004, p. 900))*

We will distinguish five elements of socio-technical systems, as they have a determining role in the diffusion of eco-design practices:

companies, state authorities, users, cultural meanings and technical knowledge.

The first element of the socio-technical systems that counts in the diffusion of eco-design practices is the companies that produce the artifacts: they can design or improve goods and services through eco-design. At the beginning of the chapter, we mentioned the figures from ADEME: in 2010, nearly a third of French companies declared that they were engaged in eco-design. They innovate in relation with other elements of the system such as universities, which bring scientific and technological knowledge and training, companies, which provide tools and machines, or distributors, which bring a network, a market or infrastructures.

The second important element of the socio-technical systems in the diffusion of eco-design methods is the States. These actors are the source of rules by issuing regulations, principles and European directives, providing technical assistance and defining green procurement policies. The eco-design methods allow us to answer the European policy of prevention of wastes since 2008.

The European Union and the French government, through public policies, establish regulations (environmental taxes, technological standards, tradable emission permits, etc.) that aim to encourage companies to deploy strategies that are more favorable to the environment (Lupton 2011). Regulatory requirements (recyclability, prohibition or limitation of hazardous substances) have multiplied since the beginning of the 1990s. They have guided the research and development projects of companies (Aggeri et al. 2005, p. 158). The aim is to encourage eco-design methods, which limit the production of waste during the manufacturing and use phases in particular.

The economic literature has examined the link between regulation and environmental innovation. From the point of view of firms, regulations impose additional costs, which may affect their competitiveness. Porter and Van der Linde (1995), through case studies, point out that pollution is often a loss of resources and that a reduction in pollution would improve productivity. Yet, environmental innovations are characterized by a double externality, which increases the costs supported by the firm and reduces incentives to invest (Rennings 2000; Galliano and Nadel 2013). Regulatory policies are then necessary to foster environmental innovations (Horbach 2008; Belin et al. 2011; Groff and Nguyen-Thi 2012; Horbach et al. 2012).

In a recent study, we show that regulations have significant and positive effects on profit levels of firms in Europe and Canada adopting eco-design (Vernier et al. 2022). They have a visible effect in the European case, with a regulatory context that encourages eco-design, unlike in Canada, where public policies are less constraining, and eco-design approaches are more often voluntary.

In addition to regulation, the European Commission is a source of norms for elements of socio-technical systems when it publishes principles and guidelines. A principle is defined by Godard (2015, p. 268) as an "abstract statement that aims to guide the resolution of a concrete problem". It allows clencher a change by indicating the rules and criteria to be followed for European policies. Thus, the polluter-pays principle gives meaning to public action. In particular, it has led to changes in tax policies on polluting activities (see Chapter 1). Europe has also issued directives that have modified strategies for packaging, waste electrical and electronic equipment, the use of hazardous substances and chemical substances. It thus encourages companies to design products or services that are more environmentally friendly (see Chapter 1).

The French government can provide financial and technical support for eco-design approaches through regional authorities such as chambers of commerce and also through its agencies. Thus, ADEME (*Agence de l'environnement et de la maîtrise de l'énergie*, the French Agency for the Environment and Energy Management), created in 1990, offers its expertise and advice and helps finance research projects in the fields of energy efficiency and renewable energies, waste management and soil conservation, air quality and noise abatement. This agency notably supports companies with a project based on eco-design. We can mention the company Douze Cycles, which manufactures cargo bikes. In 2019, ADEME supported the company to carry out a life cycle analysis of its flagship model[2] in collaboration with CETIM (*Centre d'études techniques de l'industrie mécanique*)[3].

The third element of the socio-technical systems that counts in the diffusion of eco-design practices is the users. In the first chapter, we analyzed the growing importance of the environmental issue in social representations in the 20th century. A growing number of consumers are choosing products that highlight their environmental advantage. For example, in 2020, 56% of European consumers considered the environmental impact of their purchases at least once, and 23% considered the environmental impact of most or all of their purchases (Commission européenne 2020). Similarly, purchases for energy-saving light bulbs and bulk food purchases have increased in France over 10 years (Table 3.1). This second indicator means that consumers are willing to change their purchasing habits by reducing or even eliminating packaging when purchasing. The growth in the consumption of organic products since the 1970s is symptomatic of changes in consumer behavior. It expresses a rejection by some consumers of the dominant agricultural production model based on industrialized agriculture, the capitalist exploitation model and its negative externalities (Zaidi-Chtourou and Vernier 2017). Between 1998 and 2018, the proportion of French people declaring that they have purchased at least one product from organic agriculture increased from 44% to 70% (CRÉDOC 2019). In 2021, the share of consumption of organic products in food represents 6.6% of food purchases in France according to the Agence bio[4].

Companies and governments are also users. They can add environmental criteria to their calls for tender and eco-designed

2 See: https://infos.ademe.fr/magazine-juillet-aout-2022/dossier/des-velos-cargos-encore-plus-vertueux/.

3 CETIM (*Centre d'études techniques de l'industrie mécanique*) is a technical center bringing together the public administration and industrialists, under the supervision of the French Federation of Mechanical Industries [Online]. Available at: https://www.cetim.fr/.

4 See: https://www.agencebio.org/observatoire-de-la-consommation-bio/ [Accessed July 20, 2022].

products. As principals, companies mobilize their suppliers of resources and equipment for the manufacture of eco-designed products. Aggeri et al. (2006) show that in the case of the automotive sector, suppliers must anticipate by developing eco-design tools for their industry in order to meet the requirements of their clients. The green purchasing policies of government agencies, companies with the ISO 14001 standard, architects within the framework of high environmental quality approaches or even mass retailing under the pressure of its customers, are also guiding companies' choices for eco-design approaches. We will see later that these approaches make it possible to meet the requirements of the ecolabels, which are increasingly used as proof of environmental approaches.

	Percentage of respondents who reported purchasing[5]	
	1998	**2018**
Energy-saving bulbs	47%	78%
Food products in bulk (pasta, rice, etc.)	32%	47%

Table 3.1. *Purchase of everyday consumer products with a sustainable purpose in France, 1998–2018 (CRÉDOC 2018)*

The fourth element of the system includes the cultural meanings. They are based on values related to culture and symbols. They guide consumption. Thus, users appropriate technologies, in particular through the values they convey (Geels 2004).

Since World War II, social representations have become increasingly permeable to environmental issues. In Chapter 2, we studied the way in which industrialists have progressively integrated into their production methods the modalities of reducing their impact on the environment. The current debates on the ecological transition reinforce the place of environmental values in the media and cultural field. The eco-design approach requires the company (management

5 The exact wording of the question is in 1998, "Have you ever purchased…?", and in 2018, "In the last six months, have you purchased…?".

and employees) to appropriate these values. Box 3.1 describes an experiment conducted by a design agency to help a company adopt environmental values in its production in a sustainable design in order to challenge traditional approaches. This experiment aims to break with technological routines called "standards" by Frédéric Cadet. The new concepts developed by the participants can be considered as niches paving the way to a new socio-technical regime, in this case that of the vegetal city.

Agence écoDesign offers manufacturers creative days entitled "If tomorrow we had no steel or energy". The principle is to design project scenarios based on several hypotheses that call into question traditional design methods. This reflection is based on sustainable design.

With the urban furniture company Guyon, this led to an urban architecture project that was half urban and half terrestrial. The company changed its strategic focus from competition and market share in its field of expertise. Through this approach, Agence écoDesign wishes to shift the focus of the reflection and modify the strategic objectives with a prism based on environmental and social conditions in 2030. The company has created utopias and dystopias by imagining stories of the future.

For Frédéric Cadet, director of Agence écoDesign:

> The result is the realization of a step aside, without losing the company's know-how, but with a focus on the financial validity of the market and a real differentiation on this market. Here eco-design is, initially, and before life cycle analyses, a real opportunity to dare and demystify the constraints that imprison business leaders in standard methods that do not question the company's strategy or its traditional approach to the market.

In the first step (Figure 3.2), the company participants created scenarios based on seven hypotheses (green frameworks) jointly identified during the creativity session: materials/resources, climate, green spaces, demographics, city, society and new uses. They were able to project themselves in 2030 to propose urban furniture more adapted to the new environmental and social conditions. They developed utopias (such as the "greening of cities") and dystopias (such as "more electricity"). Some scenarios were selected as strong concerns for the company (red frameworks).

In the second step (Figure 3.3), the agency developed a unified concept combining hypersufficiency and hyperlocal that fits in with the vegetal city. In the end, the company proposed a concept with 70% vegetal city, 20% sufficiency city and 10% hyperlocal. In each part, the sponsor can put parts of industrial property (here create a specific part which will be the pure know-how of the company, part which will be sold) and also locally order parts of wood. The sponsor chooses their local supplier.

At the end of this session, the company is able to create concepts to propose viable and stable foundations to build and elevate the company's solution. The company's product strategy is nourished by this soil: a founding text for future products and services. This very deep specification will be adapted to a customer demand or to a production capacity.

Box 3.1. *Creativity day (interview with Frédéric Cadet director of the Agence écoDesign, August 8, 2022)*

Finally, the fifth element of socio-technical systems includes the technical and scientific knowledge needed to deploy eco-design methods. We have seen in Chapter 2 the evolution of techniques for measuring environmental impacts. Collaboration between industrialists supports the learning process with the development and diffusion of new tools and techniques. Thus, the creation of the SETAC (Society of Environmental Toxicology and Chemistry) in 1979 aims to develop and promote tools for evaluating the environmental impacts of a technique or activity.

In France, the activity of the CREER cluster (Cluster Research: Excellence in Eco-design and Recycling) between 2007 and 2012 has the same objective. The founders are CETIM, ENSAM Chambéry (École nationale supérieure d'arts et métiers) and five companies (Renault, Steel case, Areva T&D, Plastic Omnium, Veolia Environnement and Groupe SEB). This collaboration aims to develop research on eco-design and recycling, and to share knowledge and databases to facilitate the dissemination of eco-design.

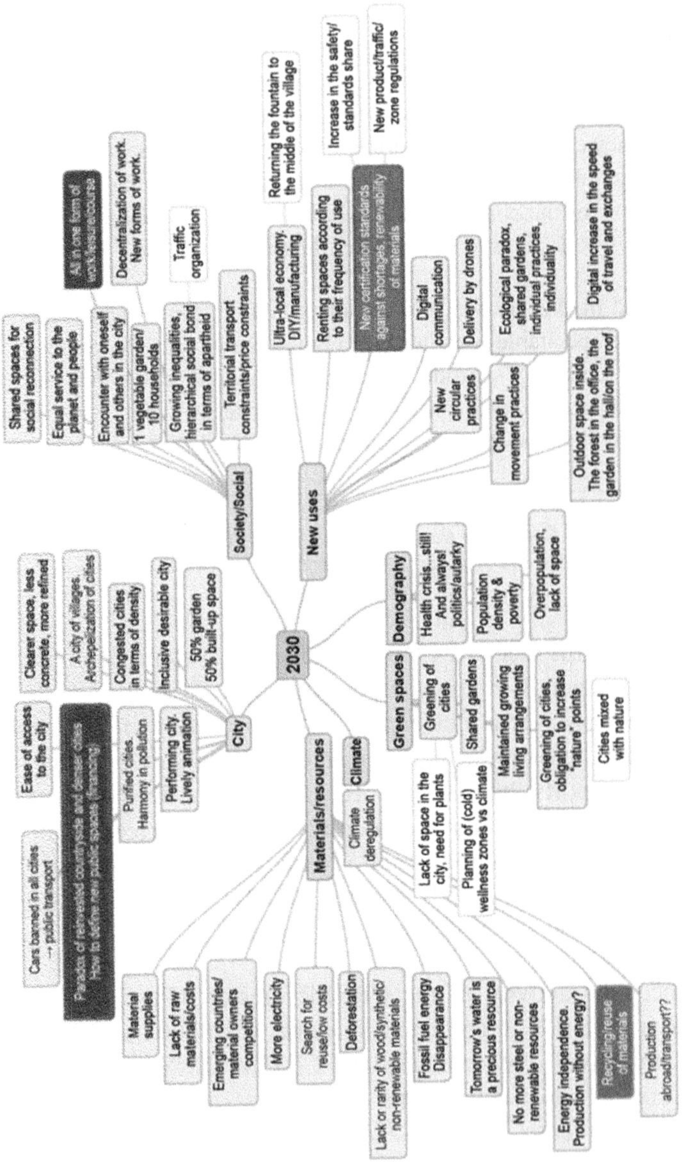

Figure 3.2. *Scenario "if tomorrow we had no more steel or energy", according to the Guyon company. For a color version of this figure, see www.iste.co.uk/vernier/ecodesign.zip*

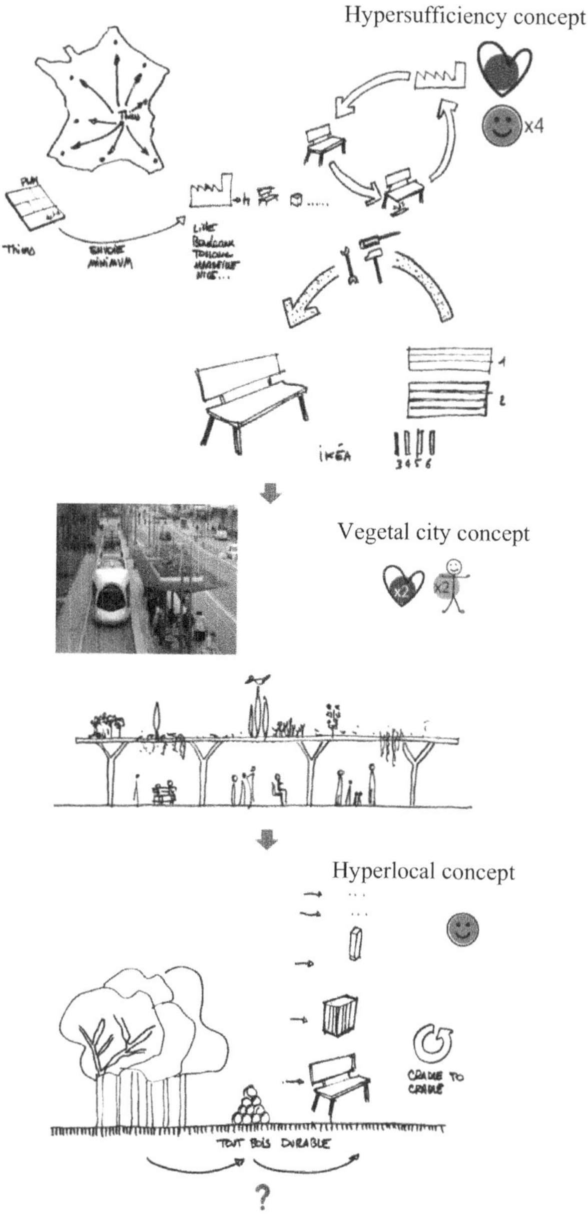

Figure 3.3. *The vegetal city (Guyon company), sketches by Philippe Moine and Frédéric Cadet. For a color version of this figure, see www.iste.co.uk/vernier/ecodesign.zip*

We saw in Chapter 2 that companies do not always have the knowledge needed to deploy eco-design (Box 2.2). Learning processes are demanding. Saidani et al. (2019) mention the lack of environmental knowledge among designers. Yet, environmental issues are complex and multidimensional. Identifying these issues influences the creativity process necessary for eco-design. The objective of LCA is to have a clear vision of the environmental impacts related to these products. The analysis enabled by LCA will be limited, as it is difficult, on the one hand, to identify all impacts and, on the other hand, to satisfy all selected criteria (Debref 2018). Margni et al. (2016) give some examples of challenges that may arise: the price of the recycled material (too high, it discourages the industrialist), the business model of the trade of recycled materials, the access to the distributors of these materials. In the case of an eco-designed product using recyclable materials, it may be difficult to anticipate the environmental consequences of recycling in terms of efficiency and the distance to collect the materials.

Vincent-Sweet (2012) and Saidani et al. (2019) highlight the complexity associated with the use of LCA. Regarding the constitution of the databases needed during the evaluation phase, obtaining relevant data is demanding due to the rapidity of technical changes, the ability to take into account criteria related to time, space and the characteristics of the territories where the resources are extracted or used (Debref 2018). Similarly, the time required to develop LCA-related tools (Saidani et al. 2019) implies that analyses may not be up-to-date.

3.2.2. *The relationships between the elements of the socio-technical systems*

Within socio-technical systems, the relationships between organizations can support eco-design approaches.

Relationships between the State, companies and professional associations encourage innovation practices based on eco-design. We mentioned earlier the positive role played by ADEME, a public agency that financially and technically supports initiatives related to eco-design. In section 3.2.1, we cited the example of the company Douze Cycles, supported by ADEME and CETIM in 2019. We also have the example of the collaboration in 2008 between a public engineering school, the State, CETIM and a company. For example, Bourgeat, a manufacturer of machines for the food industry, developed an eco-design approach for a temperature maintenance cabinet to replace a product that had existed for 15 years in the company (Berneman et al. 2009). A doctoral student from ENSAM conducted the development studies for the new product as part of his thesis, with partial funding from ADEME and in collaboration with CETIM.

The cluster *Pôle national éco-conception et management du cycle de vie* is an example of a network of industrialists and public players supporting eco-design initiatives in France and abroad (Box 3.2).

> The *Pôle national éco-conception et management du cycle de vie* is an association created in 2009 bringing together manufacturers, the Chamber of Commerce and Industry of Lyon and the Saint-Etienne Roanne metropolis. Originally, in 2005, a working group dedicated to eco-design was set up within the Chamber of Commerce and Industry of Saint-Étienne, a city known for its industrial design, notably with the École supérieure d'art et design founded in 1803. The objectives of the Pôle national éco-conception et management du cycle de vie are to raise awareness of the eco-design method among companies and to support them. It thus offers diagnostics and training.
>
> The expertise of the eco-design cluster is recognized on a national and international scale within organizations such as ADEME, the Ministry of the Environment, AFNOR, ISO and the UN Environment. The cluster is also active on a European scale within the ENEC (European Network of Eco-design Centres).

Box 3.2. *The creation of the Pôle éco-conception: an example of coordination between actors*

Collaboration between industrialists also leads to the development of eco-design approaches. Between 2007 and 2012, the CREER cluster enabled knowledge sharing. This cluster highlights the cooperation between Groupe SEB and Plastic Omnium and between Steelcase and Renault on the method of disassembling bumpers by tearing off seats[6].

Other networks support eco-design by bringing together universities, industrialists, public agencies or associations (Table 3.2).

	Network actors
GDS EcoInfo	Engineers, researchers and students.
EcoSD	Researchers, industrialists, ADEME.
ENEC	European associations of designers and industrialists supporting eco-design and in connection with the States.
Institut du numérique responsable	Actors of the civil society, of the social and solidarity economy, teachers, researchers, public, private and associative actors.
Pôle national éco-conception	Industrialists and chambers of commerce and industry.
SCORELCA	Industrialists, institutional actors and scientists.

Table 3.2. *Networks of actors around eco-design (from Bellini and Janin (2019))*

The literature shows that direct relationships with suppliers and customers can promote environmental practices in general (Greeno and Robinson 1992; Rennings and Rammer 2009). We have analyzed waste strategies in companies (Tran Dieu and Vernier 2017) based on a quantitative survey of 404 French SMEs. The results show that relationships with suppliers and customers have a significant effect on the decision to develop a prevention strategy rather than a management strategy. While the latter is a strict application of regulatory obligations, the former goes further by allowing waste to be reduced at source. In a study conducted for ADEME on the same sample (Berneman and Vernier 2014), we found that waste actions were based first on staff awareness, on actions with customers

6 *Actu-environment* [Online]. Available at: https://www.actu-environ nement.com/ae/news/cluster_creer_ecodesign_recycling_2529.php4.

(reducing outgoing packaging or containers) and suppliers (reducing incoming packaging or containers) and also on eco-design (Table 3.3).

Actions between a company and its partners lead to eco-design in nearly 20% of cases according to Table 3.3. These collaborations have been standardized with life cycle management approaches (Aggeri et al. 2005; Abrassart 2011). The eco-design method also requires building a team in the company (Box 2.2). It also relies on collaborations within and outside the company:

> Implementing an eco-design approach in a company requires the involvement of various professionals. Those in charge of the project will have to call upon several internal and external experts, or at least take them into account. This multidisciplinary approach will help establish a common vision of eco-design within the company, mobilize the various players involved around a structuring project and thus facilitate the achievement of objectives within a reasonable timeframe (Pôle éco-conception 2007, p. 33).

Internally, collaborations between employees can be the source of innovations (Tremblay 2014). Nearly 80% of companies mention employee awareness as one of their waste prevention actions (Table 3.3). As for eco-design, internal collaborations are formalized with the development of product-oriented environmental management systems (Aggeri et al. 2005; Abrassart 2011). The example of Steelcase, a world leader in office furniture, shows the importance of internal collaborations. The company's eco-design manager formed a project team in 2008 to design an office chair. The team was made up of people with different profiles, such as a commodity buyer, a logistics specialist, a designer and a commercial representative. The eco-design approach was based on an LCA. Nevertheless, the exchanges within this team allowed the identification of solutions to further reduce the environmental impact. For example, during the transport phase, a large proportion of the boxes containing the seats were empty and took up more space than necessary. Reducing the boxes saved the company "50% of space, which equated to a saving of

100,000 liters of fuel and a reduction of 345 tons of CO_2 emissions per year" (Berneman et al. 2013, p. 78).

	Number of citations	% observation
Staff awareness	210	80.5%
Optimization or reduction of the use of raw materials	169	64.8%
Reduction of outgoing packaging or containers	126	48.3%
Reduction of incoming packaging or containers	117	44.8%
New production process	79	30.3%
Eco-design (LCA)	51	19.5%
Green technologies	49	18.8%
Other actions (reduction of paper consumption, limitation of single-use products, etc.)	6	2.3%
Other (sorting, recycling, labeling, etc.)	16	6.1%
TOTAL	263	100

Table 3.3. *Waste prevention actions (from Berneman and Vernier (2014))*

Eco-design methods can lead to new relationships between the manufacturer and the users. For example, Réseau-Bureautique in Montreal offers eco-designed office furniture: it buys second-hand office furniture, redesigns it with high-end materials and resells it, mainly to companies. Working with second-hand products, it had previously encountered difficulties in its market. In 2008, it reported having an 80% repeat customer rate (Berneman et al. 2013). A second example is that of Nature Cos, a manufacturer of make-up products. The eco-design approach is based on organic agriculture certification and the sale of makeup in refillable and recyclable boxes. The relationship with the end-user is changing. In the case of lipstick, they can reuse the metal tube and buy only the refill (Berneman et al. 2013). The relationship between the manufacturer, distributor and end-user is thus strengthened.

3.2.3. *Coordination by institutions*

The activities of the elements of a socio-technical system are ordered and structured by three types of institutions: regulatory, normative and cultural.

Regulatory institutions include laws and technical standards. We previously analyzed laws enacted by the European Union and the French state to incentivize firms to develop environmental innovations. Public authorities are a key support with laws, as well as taxes and technical standards (Debref 2016). Thus, European regulations, directives and principles encourage companies to modify their production processes: the increasing costs of treating waste generated during the manufacturing process and at the end of the product's life should stimulate the search for production methods that have less impact on the environment. Eco-design is then a technical response that makes it possible to reduce end-of-life waste by improving the reuse, recovery or recycling of the components of a product to avoid paying additional costs. It will guide research and development projects (Aggeri et al. 2005).

Technical standards and environmental labels also contribute to the structuring of environmental approaches in companies. The standard agencies and professional associations in which companies exchange are at the origin of these standards and labels. In Chapter 2, we analyzed the gradual structuring of manufacturers' approaches to reducing the environmental impact of products. Several standards can be cited (Bellini and Janin 2019):

– ISO/TR 14062 (2002) "Environmental management – integrating environmental aspects into product design and development." It presents the general principles for integrating the environment into the activity, in particular at the design and development stages.

– ISO 14006 "Environmental management systems – Guidelines for incorporating eco-design." It supports the eco-design approach in the framework of an environmental management system. It can lead to changes in relationship with the company's partners in order to limit its environmental impact. For example, suppliers must respect environmental criteria.

– NF X 30-264 "Environmental management – help in setting up an eco-design approach, 2013." It accompanies the eco-design method within the company.

They refer to LCA as a tool for assessing impacts. It is explained by the ISO 14040/14044 standards (Pôle éco-conception 2017).

Industrialists have also developed standards that make it possible to integrate the eco-design approach, taking into account their sectoral specificities. Standards exist in the automotive, mechanical, electrical/electronic, construction, packaging and plastics sectors (Bellini and Janin 2019). The European Commission's official eco-label also influences corporate strategies. Created in 1992, it aims to inform users and identify environmental innovations on products. The environmental criteria associated with the ecolabel are defined by several actors: manufacturers, employers' organizations, trade unions, distributors, importers, environmental protection groups and consumer organizations. The number of products bearing the European Ecolabel has almost quadrupled between 2010 (21,301) and 2021 (89,357). The largest product categories are indoor and outdoor paints and varnishes (39%), tissue paper and paper products (17%), textiles (9%) and hard coatings (9%), according to European Commission data[7].

Accreditation encourages companies to develop environmental strategies. Thus, we have been able to show that among French SMEs, the probability of developing actions to prevent waste is significantly higher among companies with environmental accreditations (Tran and Vernier 2017).

7 See: https://environment.ec.europa.eu/topics/circular-economy/eu-ecolabel-home/business/ecolabel-facts-and-figures_en [Accessed August 23, 2022].

Second type of institutions: normative institutions, which define values, norms and expectations about the role of actors. They constitute a social obligation. Several social pressures explain the growing importance given to environmental innovations. We have mentioned the growing consumer demand for green products and public policies on green procurement. In the normative field, CSR (corporate social responsibility) is also a new rule that companies must respect in order to be credible with their audiences: customers, investors, shareholders, associations, media, etc. CSR strategies appeared in Europe in the 1990s under the terms of ethical companies, citizen companies and then responsible businesses (Capron and Quairel-Lanoizelée 2016). Social pressure on companies is increasing, as they are considered responsible for environmental degradation.

The actors of the system contribute to the definition of normative institutions based on CSR. On the one hand, the European Commission published a Green Paper in 2001 to promote a European normative framework for CSR (Commission européenne 2001b). In particular, it advocated self-regulation by companies through the implementation of codes of conduct or ethical charters. On the other hand, companies are deploying CSR strategies voluntarily or following pressure from associations supported by consumers, particularly in the textile and sports footwear sectors. They take collective action to define standards and rules in a sector. In this way, they establish the conditions for competition (Quairel-Lanoizelée 2011).

The eco-design approach leads to innovations that are consistent with CSR strategies. Some companies are changing their product strategies to limit their environmental impact. For example, the American firm Patagonia is emblematic of the companies that are succeeding in gradually converting their production methods. It will be the first to offer polar jackets made from recycled plastics. In France, Lafuma, a manufacturer of leisure and outdoor sports items, has been developing a range of eco-designed products since 2000. It starts with a backpack, and will continue with eco-designed walking shoes, jeans, T-shirts and camping furniture (Vernier 2010b). Other companies are created on the very principle of reducing environmental impact. During our study on eco-design in 2008, we met companies whose project was based on environmental innovation (the French companies Nature Cos, Plas Eco, Encre Station and Urban Cycle, the Canadian company Cascades Papiers Fins) or on eco-design (the French company Lobial and the Canadian companies Plastiques Cascades and Jouets Boom) (Berneman et al. 2009b). Their approaches are thus part of CSR.

To demonstrate the credibility of their statements, companies deploy codes of conduct, ethical charters and social and environmental certifications. Unlike environmental standards and labels, which are regulatory institutions because they are based on strict procedures, these are less controlled approaches, such as self-declarations. We will return to the different forms of communication of companies' environmental commitments in section 3.3. One example is Korhani, a carpet manufacturer in Canada that changed its carpet design from chemical-based latex to natural latex in the late 2000s. The company based its communication on a self-declared environmental "eco-product partner" (Berneman et al. 2009b) to facilitate the commercialization of this new carpet and to strengthen the company's environmental legitimacy in the market.

The third type of institution is the cultural and cognitive institutions, which can be paradigms, classifications, beliefs. They constitute a framework in which a meaning is given. Since the 1970s, individuals have become increasingly sensitive to environmental degradations and are seeking information. The media disseminate knowledge on this subject, particularly the cinema. We can mention the release of Roland Emmerich's 2004 film *The Day After Tomorrow*, which depicts catastrophic climate events, and Davis Guggenheim's 2006 documentary film *An Inconvenient Truth* about climate change.

In the industrial field, we analyzed in Chapter 2 how knowledge on environmental analysis has developed to lead, for example, to the establishment of technical standards on eco-design management and LCA. We have already mentioned the Pôle national éco-conception and the CREER cluster. The creation of the former responds to the demand for training and support in very technical approaches. The objective of the latter is to share information and build the databases needed for life cycle analyses. Some eco-design training courses are being developed in engineering schools such as ENSAM.

Systemic analysis has shown us the "capacity of actors in the environment to modify and adapt their behavior according to the transformations of their environment" (Coppin 2002, p. 34). Organizations develop learning capacities and take initiatives that promote eco-design. The systemic approach makes it possible to identify how eco-design can promote environmental innovations. The elements that support eco-design approaches are the States (through regulations and financial and technical actions), users of green products, cultural meanings and technical knowledge. Close relationships are created between organizations in the form of networks of companies, collaborations between public agencies and industry. Finally, regulatory, normative and cultural institutions structure the socio-technical systems.

3.3. Visible aspects of eco-design practices in socio-technical systems

The dynamics of innovations introduced by eco-design practices in socio-technical systems can be apprehended by two aspects: the economic results and the communication of companies.

3.3.1. *The economic results of companies: is it essential to "get" eco-design approaches accepted?*

Companies must achieve sufficient economic results to maintain or even increase their market position. The research of economic performance is a social obligation in socio-technical systems. The acceptance of eco-design approaches requires the achievement of satisfactory economic results. In other words, they are feasible as long as they are consistent with normative institutions, supporting market values in socio-technical systems.

The academic literature shows that environmental innovations are compatible with economic performance (Ambec and Lanoie 2008; Díaz-García 2015). The profitability of eco-design can be the result of lower costs, higher revenues or other benefits (Plouffe et al. 2011). Costs can decrease due to the use of recycled materials, better use of raw materials, improvements in logistics and energy savings. Most often, these cost reductions are the result of changes in one or more stages of the product's life cycle.

Revenues can increase as eco-designed products meet consumer expectations. We have also seen that for a growing number of private companies as well as for governments, the environment is a purchasing criterion. Eco-design can also lead to a simplification of the product and to a longer life span; the product is thus differentiated on its market. We have underlined that eco-design can lead to the functional economy. The company then sells a service and no longer a product, which strengthens customer loyalty. Finally, users can benefit from a reduction in the costs of using the product, which increases their loyalty (Plouffe et al. 2011).

In addition to the financial gains, eco-design can improve the company's image with its customers, as well as with investors or environmental associations, for example. It can also encourage creativity or reinforce innovation capacities. Moreover, it can better mobilize employees who are sensitive to environmental issues (Plouffe et al. 2011).

Several empirical studies demonstrate the gains of eco-design: the analysis of a government initiative in SMEs in the Netherlands in 1997, the design of an eco-designed chair in Mexico in 1999, the review by Johansson et al. (2001) of 11 eco-design cases in the electrical and electronics sectors and which remains a landmark study on eco-design, and analyses in the electronics industry in Europe in 2001 and in the printing works in 2003 (Plouffe et al. 2011).

Finally, several reports published by governments such as Environment Australia in 2001 and the European Commission in 2014 (O'Brien et al. 2014) or associations such as the WBCSD in 2000 and the BSR[8] in 2003 highlight the gains associated with eco-design approaches (Vernier et al. 2022).

We have conducted two studies to confirm the profitability of eco-design and to identify its determinant factors. In the first qualitative study, we conducted 15 interviews in France and the same number in Canada with companies that had put an eco-designed product on the market for at least one year in 2008 (Plouffe et al. 2011). For each interview, we identified the origin of the cost reductions and the increase in revenues and other benefits. Although this is an exploratory study conducted on a small sample, the results show that eco-design approaches are compatible with a good financial situation. More than half of the companies (17) mentioned a decrease in costs and a very large majority (26) an increase in their revenues. The impact on profits was measured by comparing the unit margin of the eco-designed product to that of a conventionally manufactured product: 11 companies reported a higher margin, 16 a similar margin and three a reduced margin. In this study, French firms appear to be more profitable than Canadian firms. The more profitable companies are also differentiated by a more advanced approach to LCA (measured by the number of life cycle stages retained for the eco-designed product). Thus, the ability to conduct a more appropriate LCA would promote profits. Finally, for a little less than a third of the sample (12 companies), eco-design is a personal and voluntary approach to reduce environmental impact. Normative institutions such as CSR can encourage this choice and influence company strategies.

In a second study, based on a larger sample (119 companies), we surveyed European and Canadian firms (Vernier et al. 2022). The first result confirms the trends observed in 2008: the majority of firms are making the same or more profits – 45% of firms report rising profits, 51% the same profits and 4% falling profits. The second result is the statistical analysis of the decisive actions of profitability based on a probit regression model. For the whole sample, the results show that

8 WBCSD (World Business Council for Sustainable Development) and BSR (Business for Social Responsibility) are business associations founded in the 1990s following the 1992 Rio Summit [Online]. Available at: https://www.wbcsd.org/Overview/Our-history;https://www.bsr.org/en/about/story.

regulatory motivations, customer pressure, the use of eco-design tools and a thorough LCA process significantly and positively explain higher profit levels. Nevertheless, an analysis by country shows that in Europe customer pressure and regulations have a significant and positive impact and subsidies have a negative impact on profitability, while in Canada only the use of eco-design tools has a significant and positive effect. Three types of explanations can be put forward. First, in Europe, public policies are based on authority and control, whereas in Canada, they were until recently based on the voluntary nature of companies in the environmental field (Wood et al. 2010). The dominant institutions that support eco-design in Europe's socio-technical systems are regulatory. In Canada, they are rather prescriptive: it is the social pressure in favor of environmental approaches that is decisive and allows more profit. Second, the positive effect of customer pressure in Europe may mean that the cultural and cognitive institutions in favor of the environment are stronger than in Canada, as Europeans are more sensitive to environmental issues. Third, the positive impact of eco-design tools in Canada may reflect the importance of technical standards. They promote higher profits, while Canadian laws are not very present in the field of eco-design.

3.3.2. *Communication: a tool to support the diffusion of eco-design practices*

Some companies communicate on their environmental approach in order to differentiate themselves on the market and to be more visible to potential buyers. For the consumer, the quality of a product is often unverifiable, which justifies the use of eco-labels or standards to guarantee this quality (Abrassart and Aggeri 2002, p. 4). They encourage eco-design approaches, based on LCAs, because they provide proof of the environmental approach. They thus participate in the construction of institutions supporting eco-design: on the one hand, regulatory institutions, because communication is based on standards that are often technical, and on the other hand, normative institutions that support the diffusion of environmental values in companies and among individuals. Thus, more than half of European consumers have taken into account the environmental impact of their purchases at least once. In the case of cleaning products, 78% want to know the environmental impact of their cleaning products[9].

Put simply, the term ecolabel refers to any environmental labeling or information about the environmental characteristics of a product that is provided with it. It is a means of informing the consumer (Aggeri et al. 2005). The labels are therefore numerous, and the user is confronted with a multitude of signs that they do not always understand. Moreover, their development is based on the voluntary nature of companies. Some will follow the criteria of the European ecolabel to affix this sign, while others will develop their own logo. The methods of communicating the merits of the eco-label are standardized by the ISO standard 14020, which distinguishes between ecolabels, self-declarations and ecoprofiles (Bellini and Janin 2019):

– Type I ecolabels: these are so-called official ecolabels, which have been the subject of a reference system negotiated between interested parties (Box 3.3). Examples include the NF Environnement brand, the European ecolabel, the Nordic swan and the German blue angel.

> The procedures for obtaining the ecolabels aim to complete the criteria set out in the specifications, defined by a committee representing the sector's stakeholders. There are therefore as many ecolabels as there are different products. For example, the European Ecolabel standard includes representatives from sectors such as industry, environmental groups and consumer organizations. They are the result of a multi-criteria approach to the environment and take into account the entire life cycle of products. Three ecolabels are known in Europe: the European ecolabel, the German Blue Angel and the Nordic Swan. In France, there is also the NF Environnement.

Box 3.3. *Official ecolabels (Vernier 2010b)*

9 78% of respondents agree or somewhat agree with the following statement: "I would like to know the impact of my cleaning products on the environment" (GreenFlex 2021).

– Type II ecolabels: these ecolabels provide quick and partial information, which is communicated under the sole responsibility of the person who puts the product on the market. These include, for example, the words "recyclable" or "biodegradable".

– Type III ecolabels or ecoprofiles: the information is more complex and complete. These indicators consist of a transcription of the results of the LCA of a product. One example is the Environmental Product Declaration (EPD), which may be requested from suppliers. The information provided is then checked by an accredited body.

The conditions for using the term "eco-designed" were defined in the "Practical Guide to Environmental Claims for Professionals and Consumers" published by the French National Consumer Council (*Conseil national de la consommation*, CNC) in 2012 and updated in 2019.

When using the term "eco-designed product", the company must be able to "provide relevant, significant, verifiable and concrete elements demonstrating that it has implemented an eco-design approach". The information that must be found on the product where this mention is applied are:

– A definition of eco-design (from the directive no. 2009/125 or from the ISO standard 14062).

– Details on what is eco-designed: the product, the packaging or a component. Otherwise, the term refers to the product as it is marketed, that is, including packaging.

– The main environmental characteristics of the product and/or its packaging.

– The nature and, if possible, the extent of the reductions in environmental impacts resulting from the eco-design approach (this information can be included in another appropriate medium, such as a website).

Box 3.4. *Communication on eco-design according to ADEME*[10]

10 See: https://www.ademe.fr/entreprises-monde-agricole/organiser-demarche-environnementale/dossier/ecodesigning-products/issues-ecodesign-benefits-business-environmentmie-lenvironnement [Accessed August 23, 2022].

ADEME distinguishes between so-called eco-designed products and products bearing the European ecolabel. In the first case, the user cannot know the exact level of environmental requirements, whereas in the second case, the official ecolabel is based on an LCA of the product (Box 3.4). ADEME thus supports technocentric eco-design approaches by promoting their advantages in terms of communication.

Finally, the food market for organic products has similarities with that of eco-designed products. Organic production aims to reduce the environmental impact of an activity; it is recognized as organic on the basis of specifications (the European organic farming label) controlled by the European Union, and it leads to changes in farmers' practices. In this market, which is growing rapidly in terms of volume and value, players must develop communication strategies to defend their legitimacy with their customers. Thus, specialized organic brands, pioneers in this market, coexist with generalist retailers that sometimes invest aggressively in this market and with significant financial resources. The former benefit from a stronger legitimacy than the latter due to their historical presence. The cultural and cognitive institutions in favor of the consumption of organic products are increasingly important even though organic food purchases remain a minority. Both specialized organic and generalist chains are seeking consumer support in order to consolidate their position in the market. Thus, general retailers are developing organic ranges, specialized departments and now offer 100% organic stores. We have shown that the historical legitimacy of organic chains was not sufficient in the face of the actions of generalist chains to position themselves on this market (Zaïdi-Chtourou and Vernier 2017). Using Suchman's (1995) grid, we identified three types of legitimization strategies by specialty chains:

– A pragmatic legitimization strategy in a market logic with the use of marketing and commercial techniques. The "organic" is then a pretext to attract the consumer.

– A strategy of moral legitimization, the specialty retailer and the consumers share common standards, beliefs and values. The retailer asks consumers to take into account its commitment to organic products by buying responsibly.

– A strategy of cognitive legitimization, the brand highlights the behaviors that are recognized as acceptable in the market and that everyone should adopt with arguments based on the health and well-being associated with organic food consumption.

In the face of fierce competition from general retailers, the specialist retailers are defending their legitimacy. We can take up the concepts of the socio-technical system in the food sector: the market here is the food distribution area. The organic chains distribute products from technological niches: agricultural production based on alternative techniques to those that consume chemical inputs and are highly mechanized. They are attracting growing interest from end-users and therefore from general distributors. Organic agricultural production is still a minority compared to the prevailing regime of conventional agriculture. The communication strategies of specialized retailers are based on organic food as a new consumer standard in the socio-technical system in order to defend their historical legitimacy in the face of general retailers, who are most often associated with conventional agriculture.

3.4. Conclusion

We are seeing a dynamic of eco-design practices in socio-technical systems. The elements composing them (companies, states, users, cultural meanings and technical knowledge) contribute to the development of eco-design. Institutions coordinate these activities in the form of laws, norms, values and beliefs and orient the elements of the system in favor of eco-design. Regulatory institutions have strongly encouraged its diffusion through laws and technical standards that explicitly support pollution reduction at the source. On the contrary, if the normative (CSR) and cultural-cognitive (media) institutions have supported this extension, it is rather in favor of all environmental innovations. We have noted the economic performance associated with eco-design approaches and the importance of technocentric eco-design, which, along with LCA, provides proof of the environmental approach.

The technocentric approach, driven by LCA, dominates (Petit 2015; Debref 2016, 2018). Innovations do not challenge the technological regime and are most often incremental (Abrassart and Aggeri 2008).

At the end of our analysis, it appears that:

– eco-design is supported by regulatory institutions, with public regulations and technical standards often based on LCAs;

– eco-design is at the heart of relations between organizations that form networks to disseminate it (Pôle éco-conception), to share databases (CREER cluster) or to conduct an LCA (collaboration between ADEME and companies such as Bourgeat);

– the study conducted on the profitability of companies shows that motivations related to regulations and customer pressure, the use of eco-design tools and a thorough LCA process significantly and positively explain higher profit levels;

– eco-design is supported by communication through official ecolabels, labels supported by the European Union and based on LCA.

In his thesis, Abrassart (2011) shows the importance of LCA as an analytical tool structuring the training of engineers in eco-design and thus their practices in companies. LCA is at the heart of one of the elements of socio-technical systems that are technical knowledge.

We can thus wonder about the dynamics of eco-design, which is undermined by the technocentric approach and leaves little room for sustainable design approaches that combine environmental and social innovations. The latter can lead to radical innovations that go against the dominant technological regime in socio-technical systems. The circular economy and functional economy approaches may be a response. This is the subject of Chapter 4.

4

Questions and Perspectives on Eco-design

In the previous chapter, we noted that eco-design practices were at the origin of environmental innovations in socio-technical systems. The purpose of this chapter is to identify the limits and perspectives of eco-design within the framework of the reflection on the ecological transition. We will question, in a systemic approach, the capacity of the actors to propose sustainable modes of design and go beyond the dominant regime, in order to elaborate new rules creating the conditions of a strong sustainability.

To this end, we will begin in the first section with an assessment of technocentric eco-design, which, even though it effectively allows for the identification and reduction of the main categories of environmental impact, does not allow us to move away from the dominant technological trajectories. In a second section, we will analyze how the circular economy is an opportunity to change the logics presiding over material flows, enabling new business models that force companies to change their practices (Bocken et al. 2016; Lüdeke-Freund et al. 2018). Users, designers and manufacturers need to collaborate differently from a sustainability perspective. Nevertheless, we will see that the circular economy is not sufficient to address the ecological crisis. In a third section, we will study how sustainable design can foster new uses and we will look at the strengths of frugal innovation. Finally, in the last section, we will identify the perspectives offered by an open innovation process and the notion of common good.

4.1. Technocentric eco-design: an insufficient solution to the ecological transition

4.1.1. *The difficulty of leaving the dominant socio-technical regime*

We pointed out in Chapter 3 that niche innovations emerged with difficulty in the socio-technical regime, as the actors of the system resisted (Geels 2014). The European Commission, in the report by O'Brien et al. (2014), and the article by Debref (2016) note that environmental innovations are most often incremental: the modifications are made at the margins and remain in line with the dominant socio-technical regime. In the mid-2010s, for the European Commission, the most common approach in the environmental field was to improve energy performance according to O'Brien et al. (2014).

We also highlighted in Chapter 2 the large number of tools available in technocentric eco-design approaches. The variety is an advantage, as it allows the most relevant tool to be used in view of the product's characteristics. But this requires designers to master scientific and technological knowledge of the socio-technical system in the production sphere, which can be difficult. Thus, Knight and Jenkins (2009), based on a study of an industrial company supplying equipment to the military and security sectors, identified several obstacles to the adoption of eco-design. First limitation, tools such as LCAs, guidelines and matrix approach, are often considered to be expert tools. In particular, LCA is perceived as too complex and not very easy to practice. The second limitation is that the requirements of eco-design tools come up against the need to make a product that meets the expectations of the end customer. In the end, the company's approach results in a limited eco-design method. As a final limitation, LCA is a tool to identify present and recognized problems based on a well-defined situation (den Hollander et al. 2017); it therefore limits eco-design to improving what already exists.

This leads us to wonder about the conditions for conception. Abrassart and Aggeri (2008) showed that eco-design is most often situated within the framework of a regulated design. It is then limited to innovations at the margin within a dominant design and established technological trajectories and does not encourage innovative approaches that combine creativity and respect for the environment. We will study the possible situations built around two axes:

– The horizontal axis defines the environmental value's mode of construction. We will define it as the value associated with the non-degradation of the environment during the production process. It is estimated by the difference in environmental consequences between a production method that limits the effects on the environment and a traditional production method. The environmental value is said to be confined when the criteria and specifications can be defined at the beginning of the design process. On the contrary, it is said to be exploratory when they cannot be specified at the beginning of the process (Box 4.1).

– The vertical axis distinguishes between the selective and the systemic dimension of the eco-design process. In the first case, the process is focused on a specific environmental dimension, while in the second case, it encompasses several dimensions considered simultaneously within the framework of a systemic life cycle approach (Box 4.1).

From these two axes, Abrassart and Aggeri (2008) identify four eco-design situations (Table 4.1). In cases 1 and 3, the specifications can be defined upstream of the project; the eco-design then simply aims to improve the ecological performance of the product:

– Case 1: selective and confined eco-design responds to a situation with regulations. There is no reflection on the environmental criteria to be retained nor on the environmental value for the customer.

– Case 3: life cycle design, which is the most common situation in the literature. The design objectives are defined and the means to achieve them are relatively known and stabilized. It is the LCA that allows the quantification of inputs (raw materials, energy, etc.) and outputs (emissions, waste) at all stages of the life cycle.

Construction of the environmental value

1) Confined: the environmental value is limited to an impact criterion (see Chapter 2) such as the depletion of the ozone layer. To achieve this objective, the designer will have to eliminate about 10 substances such as methyl bromide and CFCs (chlorofluorocarbons), gases used for refrigeration and air conditioning.

2) Exploratory: the environmental value is identified as the design progresses.

Dimension of eco-design

1) Selective: only one environmental impact is considered. For example, in the case of laundry detergent manufacturing, an industrialist can exclude the use of phosphate in accordance with the regulations.

2) Systemic: several environmental impacts are considered. In 2018, Alstom conducted life cycle assessments (LCAs) on the products that make up all of its mobility solutions. In the second step, Alstom integrated consequences not covered by the LCA such as noise, landscape impact, electromagnetic emissions, fine particles or even hazardous substances and their repercussions on the environment and health (Pôle éco-conception 2021, fiche éco-initiative Alstom).

Box 4.1. *Examples illustrating eco-design typologies*

		Construction of the environmental value	
		Confined	**Exploratory**
Dimension of eco-design	**Selective**	Selective eco-design (1)	Exploratory selective eco-design (2)
	Systemic	Eco-design regulated by the life cycle (3)	Systemic innovative eco-design (4)

Table 4.1. *Eco-design situations (from Abrassart and Aggeri (2008))*

In the other two situations, the environmental issues are less framed and require more creativity from the various partners:

– Case 2: exploratory selective eco-design. The project brings several partners (recycling experts, demolition companies, waste disposal, etc.) to work together. We can think of the development of products in the household appliances sector, for which several aspects of production, and also of distribution and end-of-life must be considered upstream because of the extended producer responsibility.

– Case 4: systemic innovative eco-design. In this case, neither the possible values of the environment, nor the associated knowledge nor the organizational forms are defined in advance. It is then a question of designing more innovative solutions that may be necessary to get out of a technological trajectory that has exhausted its effects. The break is therefore significant when considering the environmental value and the technologies used. Nevertheless, the designer may be constrained in their search for solutions when scientific knowledge does not offer answers to complex issues.

Systemic innovative eco-design can lead to a change in technology trajectory in which a niche innovation emerges and changes the dominant technology regime (Geels 2014). It goes beyond the limits set by LCA. We are then in a situation in which eco-design is synonymous with creativity and respect for the environment. It can then contribute to the ecological transition.

We illustrate the difficulties of getting out of the dominating socio-technical regime with two examples from the flooring and small household appliances sectors. The first case is taken from the thesis of Debref (2014); it shows that environmental innovations come little out of the dominant technologies due to phenomena of resistance in the sector. He identifies four situations: (1) manufacturers will fight to keep PVC (petroleum-based thermoplastic polymer) as a raw material despite the existence of alternative and chlorine-free raw materials; (2) process innovations are developed, but they are based either on technologies at the end of the chain which consist of correcting existing pollution or on eco-efficiency where the search for savings takes precedence; (3) incremental innovations on products are observed, which limit the use of glue and thus facilitate recycling; (4) an old product based on an old technology (linoleum) is coming back on the market. Maintenance in the dominant technological regime comes from the fact that the actors of the socio-technical system support existing products (case 1), technologies at the end of the chain (case 2), incremental innovations (case 3) or old technologies (case 4). The change in practices is therefore limited.

The second example concerns a product eco-designed by Groupe SEB in the late 2000s (Vernier 2023). It shows that an innovation combining many advantages in the production and use phases can fail. In 2008, Groupe SEB launched an eco-designed vacuum cleaner that breaks significantly with conventional design with five main features: new materials, lower energy consumption, ease of use, durability and repairability. During the research and development phase, Groupe SEB received financial support from ADEME, the collaboration of an engineering school (ENSAM) to conduct an LCA and the support of a design firm to design the product. The launch of the vacuum cleaner on the market was not successful. On the one hand, the main criterion of choice for users is the power consumption of the vacuum cleaners, which is synonymous with suction power. Also, for manufacturers, the selling point is kilowatt-hours. On the other hand, only a minority of consumers were willing to buy a green product in the late 2010s. In 2007, 35% of consumers said they would choose a product made by a company committed to environmental protection at a higher price (5–10%) and 41% said they would hesitate to buy a conventionally manufactured product (Vernier 2023). Finally, Groupe SEB had communicated very little on the environmental characteristics of its products, even though it was one of the leaders in its market and a pioneer in eco-design. In the end, sales were insufficient and the vacuum was withdrawn from the market. It was not until 2014 that energy labels for vacuum cleaners became mandatory and the maximum allowable electrical power of a vacuum cleaner was capped at 1,600 W. In the socio-technical system built around vacuum cleaners, the elements are not favorable to environmental innovations. The eco-designed vacuum cleaner by Groupe SEB remains a niche innovation that has not been able to counter the dominant socio-technical regime.

4.1.2. *Limitations due to eco-efficiency and the rebound effect*

Eco-design makes it possible to reduce the environmental impact of a product throughout its entire life cycle thanks to a precise analysis of the effects of the product's components, from its design to its end-of-life, by the designers' decisions on manufacturing methods. Nevertheless, the expected benefit for the environment can be limited for two reasons: the search for eco-efficiency and the rebound effect.

4.1.2.1. *Eco-efficiency*

Technocentric eco-design can lead to material savings (Bourg and Papaux 2015) or improved energy performance (O'Brien et al. 2014). It thus follows the principle of eco-efficiency (Knight and Jenkins 2009). It was introduced as a pathway to sustainable development in the early 1990s, and it was at the Earth summit in Rio in 1992 that Stephan Schmidheiny and then the World Business Council for Sustainable Development[1] association would defend this principle (Lanoie and Plouffe 2008; Debref 2018). The search for eco-efficiency can be done in different ways: reduction of material intensity, energy intensity and dispersion of toxic substances, improvement of recyclability, maximization of the use of renewable resources, extending resources, extending service life and increasing service intensity.

The principle of eco-efficiency is thus compatible with better economic performance thanks to the reduced use of materials or energy and reduced losses in the process (Braungart et al. 2007). It consists of minimizing the volume, speed and toxicity of material flows without calling into question production methods (Braungart et al. 2007).

1 The WBCSD is a business association created in 1995 by Stephan Schmidheiny, mandated by the United Nations Conference on Trade and Development (UNCTAD) to disseminate the concept of sustainable development among businesses, following the 1992 Rio Summit [Online]. Available at: https://www.wbcsd.org/Overview/Our-history.

But there are several limitations to this approach. The first is that the pursuit of resource efficiency has as its primary goal less resource use per product (Braungart et al. 2007; Bocken et al. 2016; Debref 2018). Indeed, this principle is built into a view of the economy as "a linear, one-way flow of material, where raw materials are extracted from the environment, transformed into products, and sometimes disposed of" (Braungart et al. 2007, p. 1337). The search for resource efficiency does not address the problem of limited natural resources, but rather postpones the date when they will no longer be sufficient.

The second limitation concerns production methods. On the one hand, material flows in companies are unchanged, and on the other hand, the toxic impacts of industrial activities remain. Thus, Braungart et al. (2007) note that the search for eco-efficiency leads to modes of degraded recyclability where recycled materials lose their qualities. These authors then propose as a solution the cradle-to-cradle approach, which will then be integrated into the circular economy: resources of biological origin and resources of technical origin are transformed, recovered and reused while maintaining the highest possible value throughout the life cycles of the products.

For his part, Debref (2018) notes the paradox of a situation where the search for efficiency actually leads to better resource productivity and also to low material use. The question then is the limit to intensification in resource use (Debref 2018). It also considers that the modes of production are not fundamentally changed. Eco-efficiency is in fact associated with the search for technologies, methods and materials that increase the resources used for the same production.

4.1.2.2. *The rebound effect*

While eco-efficiency allows us to produce more with less, these resource savings can give rise to a "perverse" effect called the rebound effect (Arnsperger and Bourg 2016; Debref 2018; Korhonen et al. 2018a).

For example, in the automotive industry, engines are more efficient and use less fuel. But these gains are reduced because the vehicles produced are heavier (Bihouix 2014). The rebound effect is therefore a reduction in potential gains. It refers to "the increase in use or consumption of one or more products following a gain in efficiency – measured energetically, materially, temporally, etc. – in the production, supply or use of a good or service" (Bourg and Papaux 2015, p. 383). Rebound effects can be identified at the micro and macro levels (see Box 4.2). The expected gains from lower resource consumption are lower than the gains initially hoped for, in the short or long run. Over time, the pursuit of resource efficiency can lead to accelerated material flows by increasing the sale of more efficient products (Bocken et al. 2016).

At the micro-economic level

The direct effect: the improvement in energy efficiency leads to greater consumption of the good requiring the energy input. For example, a more fuel-efficient car will cost less per kilometer, thus providing an incentive to drive more.

The indirect effect: the monetary gains from lower energy consumption are spent on other goods, the production of which requires energy and the process of which can be polluting. For example, the gasoline saved over several months because of a less polluting car can be used to buy a plane ticket.

At the macro-economic level

The market effect: this is the corollary of the direct micro-economic effect on a larger scale. As the total demand for energy or resources falls, its price will fall, making some activities profitable, which in turn increases demand. For example, we can imagine a shift from rail to road freight if road transport becomes less expensive, because gasoline consumption per kilometer decreases.

The growth effect: an improvement in energy or material efficiency can positively affect growth in several ways and lead to greater energy consumption at the macro-economic scale. A more efficient sector can induce investment and thus grow to offset or overcompensate for initial gains. If this sector is central to the production system, these gains can be transmitted to all other sectors in the form of increased productivity, and thus stimulate economic growth, and thus resource and energy use. In addition, the money saved by lower spending at the macro level can be used to mobilize fallow resources in a situation of underemployment, giving rise to a Keynesian multiplier effect.

Box 4.2. *The rebound effect at the micro and macro-economic levels (after Daumas (2020), pp. 190–191)*

Daumas (2020), in a comparative analysis, shows the difficulties of measurement due to the diversity of methods and the time scale. Rebound effects could represent between 10 and 80% of initial gains (Daumas 2020, p. 194). According to Bourg and Papaux (2015, p. 383), they vary between 10 and 30% of the service-related energy consumption of certain goods such as automobiles and housing in OECD countries.

The eco-design method can contribute to rebound effects. We explained in Chapter 2 that in view of the LCA results, the company develops its product strategy. It selects its environmental strategy according to the results of the LCA on the product and the impact of this reduction on each stage (costs, materials to be used, energy required, weight of the final product) (Berneman et al. 2009a; Vernier et al. 2022). To avoid the rebound effect, the final product should not have a reduced environmental impact on one stage of the life cycle and an increased impact on another stage. We can look at its effects by life cycle stage from a macro-economic point of view. When comparing electric cars to combustion cars, environmental emissions are shifted from the use stage (lower greenhouse gas emissions in the absence of fuel) to the production stage (mineral resource consumption and toxicity increase due to power plants, the distribution network and battery production) (Margni et al. 2016).

Finally, technocentric eco-design leads to an innovation that certainly generates gains in terms of environmental impact, but these are partially reduced by the search for eco-efficiency and rebound effects.

4.2. The contribution of eco-design to the circular economy

We have defined the circular economy as the development of activities based on resource loops (of materials and energy) in the form of reuse, remanufacturing and recycling, and which are no longer discarded as wastes at the end of their life. We are interested in the business models deployed by companies engaged in circular economy practices. We will see that product design, and particularly eco-design, contributes to these practices. The notion of business model also allows us to show the important role of relationships with actors of the socio-technical system such as users, sellers and suppliers.

4.2.1. *Circular economy business models and the diffusion of eco-design*

The notion of business model is to analyze how the company creates value. It is now considered a key driver of innovation (Bocken et al. 2016). This term, dating from the 1990s, is associated with the world of the Internet and new technologies, for which future cost and profit structures had to be estimated. These companies were pioneers in their field and could not be evaluated based on their past performance (DaSilva and Trkman 2014; Masoumeh and Maryam 2020). The term business model is used to explain value creation analyzing how "the firm delivers value to its customers, attracts customers to pay for that value, and converts its payments into profit" (Teece 2010, p. 191).

A substantial literature has developed on circular economy business models (Bocken et al. 2016; Urbinati et al. 2017; Frishammar and Parida 2019; Lopez et al. 2019; Lüdeke-Freund et al. 2019; Pieroni et al. 2019; Urbinati et al. 2021). The main challenge for companies developing the circular economy is how they create and deliver value through their business models (Lüdeke-Freund et al. 2019). A key success factor is the concurrent deployment of product design and business model strategies (Bocken et al. 2016). Few companies transform a product-based business model into an outcome-based model (delivering functionality). They then have multiple business models simultaneously (Frishammar and Parida 2019). We saw the case of Établissements Cros: they continue to sell machines and also offer outcome-based services (offering cubic meters of compressed air).

Bocken et al. (2016) are particularly interesting. They analyze circular economy business models by detailing how product design contributes to them. They identify six business models that enable innovation for the circular economy by slowing down or closing resource loops[2]. These are new business models, as value creation is based on a unique way of operating, with a new organization of relations between the company and the environment in which it operates. The strategies of partners (suppliers, distributors, customers, for example) and the behavior of users are modified. We will analyze these business models by identifying their originality (proposition, creation and capture of value), the company's relationships with other actors and the type of product design in order to assess the role of eco-design (Table 4.2).

	Business model	Product design	Examples
Business model to slow down the loops	Access and performance	Eco-design Design for sustainability, reuse and/or reparability	Établissements Cros Yit
	Extension of the residual value of products	Remanufacturing	Remober Envie
	Extending the life of products	Eco-design Designed for durability and repairability	Asphalte
	Support for sufficiency	Eco-design Design for Sustainability	Patagonia
Business model to close the loops	Extension of the residual value of resources	Design from recycled materials	Ecclo
	Industrial ecology	–	Kalundborg

Table 4.2. *Circular economy business models and product design*

2 The authors excluded activities consisting solely of narrowing material flows, as they are based on the principle of eco-efficiency and do not contribute to the circular economy.

There are four business models that slow down the loops. The first model is the access and performance model, which is based on two characteristics. The first is that the producer remains the owner and the second is that it delivers a service. We find the principles of the functional economy and product-service systems. These are circular economy business models, as they reflect a new configuration of activities to enable the slowing down of material loops. The value proposition is based on the service allowing the user to benefit from the functions of the product. Regarding the creation of value, on the one hand, the producer increases their profits with products designed to last, for reuse, to be repaired or for eco-design, and on the other hand, the user reduces their costs, as they pay per use and reduces their need for material goods (Manzini and Vezzoli 2003; Bocken et al. 2016). Value capture is based on price per unit of service. This model is based on new relationships with the company's partners with converging economic interests. It is compatible with eco-design, but this is not systematic (Buclet 2014), and the design of the product can solely be oriented towards durability and reparability (Arnette et al. 2014). This model is particularly interesting because it can associate new product design mode (here eco-design) and new behaviors in the user. We gave the example of Établissements Cros, which makes compressed air available and has thus developed a new business model alongside its core business. In particular, this helps to reduce the environmental impact. In the case of a contract with a company in Grenoble[3], this solution makes it possible to limit energy consumption by maintaining a contractual ratio. A compressed air production unit is installed on the customer's premises, and Établissements Cros carries out remote monitoring on a permanent basis and regular maintenance. The Finnish manufacturer Yit also offers a car-sharing service to the residents of its Tripla complex in the suburbs of Helsinki. In this way, Yit is reducing the number of cars per resident and changing behavior. To date, nearly 200 of the 400 apartments have subscribed to this service[4]. Almost a quarter of the residents have stopped owning a car.

3 See: http://www.cros.com/fr/nos-realisations/fourniture-dair-comprime-cle-en-main-dit-au-m3?termid=30 [Accessed September 14, 2022].

4 See: https://www.yitgroup.com/en/news-repository/press-release/popular-shared-cars-at-tripla-have-been-a-factor-to-consider-when-moving-to-pasila [Accessed September 13, 2022].

And among the residents who do not own a car, nearly 80% have tested this service.

The second business model is to extend the value of products by extending their residual value through remanufacturing or collection of products. The value proposition is to transform the residual value of the product into a new product. The company creates value through product collections and collaborations with distributors, logistics specialists and relay points and then remanufacturing. It appropriates new forms of value, as the total cost decreases with the decline in material costs. The user benefits from a reconditioned product with a reduced price compared to a new product. The design is oriented towards remanufacturing. It should be noted that in Europe, regulatory institutions (WEEE and extended producer responsibility) have strongly encouraged companies in the field of electrical and electronic equipment to take back their products for reuse or recycling. In Chapter 2, we presented the company Remober, which buys back old electronic equipment, repairs it and puts it up for sale. The particular case of the Envie network, created by the Emmaus association, is also interesting. The association started a collaboration in 1989 with a first shop Darty in order to recover the products that were thrown away, repair them and sell them with a guarantee[5]. In the 2000s, the network was extended throughout France and with other partners (local authorities, other distributors) due to the WEEE directive. Envie also has the particularity of combining an environmental and a social objective, and that from the outset it has been about creating jobs for employees excluded from the labor market. In the case of Remober and Envie, the creation of collaborations is essential between the company and the specialists for collecting, sorting, repairing and selling.

A third business model is based on extending the life of products. The value proposition is based on longevity, as well as on quality and services (repair and reuse possible over time). Value creation is enabled by a product conception based on duration and a high level of service. The company appropriates the value with a high selling price covering these services. Companies are also creating business models

5 See: https://www.envie.org/decouvrir-envie/notre-histoire [Accessed September 14, 2022].

that combine sustainability and affordability for the customer. These are business models often associated with the luxury sector. However, we can take the example of the company Asphalte, which offers pre-ordered T-shirts[6] at relatively high prices. It creates value because it only produces the T-shirts once a minimum order threshold has been reached, thus reducing losses linked to unsold merchandise. Customers change their buying habits for two reasons: first, they order and have to wait several weeks before receiving their product, and second, the number of models is limited. The design can be eco-designed or solely oriented towards sustainability and repairability.

The last business model aims to support sufficiency and reduce consumption by the end-user with a non-consumerist approach to sales. The value proposition is based on sustainability and a very high level of service maintained over time. Value creation is enabled by a product designed to last and be of high quality. The company may have reduced quantitative sales targets and instead have a non-consumerist approach to its market. It captures value through the higher prices of its products. The product design can be based on eco-design or solely oriented towards sustainability. Customer relations are no longer based on sales, but on sufficiency. For example, Patagonia, an American manufacturer of sporting goods, offers high-end sporting goods. It encourages its customers to buy only what they need, repair, reuse and recycle (Bocken and Short 2016). This company is also known for its commitments to environmental approaches, including eco-design (Vernier 2013).

There are two business models for closing resource loops. The first business model is to extend the value associated with resources by extending the residual value of resources considered as waste. The value proposition is to make the product more attractive to users and lower the cost of materials and the final price. The creation and capture of value is based on the organization of a resource collection system with stores, sorting centers or consumers and then designing a product. Collaborations with these partners are essential to the sustainability of the model. The company Ecclo[7] offers clothes to the

6 See: https://www.asphalte.com/f/pages/notre-mission-f [Accessed September 14, 2022].

7 See: https://ecclo.fr/pages/ecclo-en-details [Accessed September 14, 2022].

general public made from fabrics or textile threads that have to be thrown away (surplus or manufacturing defects).

The second business model is the industrial ecology, which was studied in Chapter 2. We can call it a business model because it is based on a new organization of relations between the partners at the origin of an industrial network. The value proposition consists of reducing costs and operational risks within this network. Companies create value by pooling costs, exchanging by-products, creating new activities from waste streams and taking advantage of lower material costs. The industrial symbiosis in Kalundborg is an archetype. A power plant, an oil refinery, a biotech and pharmaceutical company, a gypsum board producer and a soil remediation company work together. They mainly exchange materials (water and solid waste) and energy (heat and steam) (Diemer 2016).

At the end of this study, the functional economy, considered here as a business model based on access and performance, must be distinguished. Indeed, it allows us, on the one hand, to replace the access to the product by the use of its functionalities. On the other hand, the eco-design method allows us to go further by evaluating and then reducing the environmental impacts associated with the product over its entire life cycle, which is modified, since the product remains the property of the company.

4.2.2. *Technocentric eco-design and the circular economy: complementary approaches*

Eco-design and circular economy have different objectives: the former aims to reduce environmental impacts, while the latter aims to eliminate wastes into materials and energy (den Hollander et al. 2017). According to Samuel Mayer (Box 2.2):

> The circular economy is based on loops to reduce waste, but the question of environmental performance is not asked.

Thus, the search for a lower environmental impact over the whole life cycle of the product (definition of eco-design approaches) can lead to a reduction of waste (characteristic of the circular economy), whereas the opposite is not always true. Indeed, waste can be reduced, but it may have a higher environmental impact due to new production processes, materials used or the transportation of materials to be recycled. For Bellini and Janin (2019), the circular economy is a solution to reduce the consumption of raw materials that does not systematically avoid increases in pollution. Eco-design then appears as a preliminary and necessary step to have a vision of the whole life cycle of the product (Bellini and Janin 2019; Bruel et al. 2019) in order to reduce the environmental impact associated with a circular economy practice. For example, Brunet Éco-aménagement, using the eco-design method, has favored the use of recycled materials and implemented waste recycling during the manufacturing stage of the housing developments. It has modified its production process with the reuse of production offcuts by partners (Berneman et al. 2009). In this case, eco-design first leads to a reduction in environmental impact, which is the primary objective, and second, contributes to the circular economy.

Eco-design and circular economy are then complementary. In section 4.2.1, we found that the type of product design, and in particular eco-design, had a decisive role in circular economy business models. According to Bruel et al. (2019), successful deployment of the circular economy requires thinking at the micro-, meso- and macro-economic levels. Eco-design is at the first level as a product design method (Bruel et al. 2019). Product design and industrial processes have a significant effect on the sourcing, consumption of resources and generation of waste over time (Mendoza et al. 2017). For example, Roux et al. (2005) estimate that the design of the product determines 90% of its cost and performance, including environmental, over the entire life cycle. In 2012, a study by the European Commission considered that the product design stage determined more than 80% of the environmental impacts over the product's entire life (Mendoza et al. 2017).

4.2.3. *Circular economy business models and relationships between actors in socio-technical systems*

Circular economy business models have economic, environmental and also social consequences, especially because they create new relationships between actors. Their analysis shows that the physical flows of materials and energy cross the boundaries of processes, organizations and sectors. The actors are numerous with interests and preferences that may differ. Governance, management and decision-making can thus be difficult (Korhonen et al. 2018b). As Manzini and Vezzoli (2003) point out about product-service systems, this is an opportunity for new relationships between partners, a new convergence of economic interests and an optimization of system resources.

We can take up the analytical grid presented in Chapter 3. Indeed, we can analyze the elements of the socio-technical system to understand the changes generated by the new circular economy business models. First, the relationships between the three spheres (production, use and distribution) in the socio-technical system are changed. Artifacts circulate under new conditions. Thus, in the business model of access and performance, the producer remains the owner. In the distribution sphere, the distributor offers a service around the product, or even disappears if the producer is in direct relation with the user. In the business model for the extension of product value if they are dual, within the repair and maintenance systems, the sphere of use becomes more important. Technical knowledge and expertise extend the value of the product. In addition, regulatory institutions support reparability. For example, from January 1, 2021, as part of the circular economy anti-waste law, the French state introduced the reparability index on five types of equipment. This encourages manufacturers to design and users to choose goods from a reparability perspective. Second, cultural meanings are changed. Indeed, products considered as waste become products to be remanufactured, renovated or recycled. They acquire an economic value. These wastes, which were supposed to leave the socio-technical system, are transformed and circulate again between the three spheres. It is the regulatory and cultural-cognitive institutions that explain the change in status of waste in the system. Third, the eco-design method contributes to the circular economy business models. It spreads by relying on the scientific and technological knowledge that are the elements of the socio-technical systems of the production sphere. Regulatory institutions, through technical standards, and normative institutions, through CSR (corporate social responsibility) and labels, support this diffusion. Eco-design methods are thus completed by design methods oriented towards reparability, durability or recyclability.

Finally, some circular economy business models constitute an opportunity in favor of the ecological transition by relying on behavioral changes at the individual level. Thus, the access and performance model reinforces the role of users (Petit 2015). The latter deliberately choose not to be owners anymore, but to access a service. For Arnsperger and Bourg (2016), even though the functional economy does not completely solve the problem of long-term resource scarcity (see section 4.2.4), it has two advantages. On the one hand, for the producer, the reduction of resource consumption makes the service more profitable. On the other hand, the product is accessible to a larger number of users, because each user uses it only when they need it.

Similarly, the industrial ecology model is often analyzed from a technical perspective (Buclet 2011). However, it can be understood on the scale of a territory, in which case the participatory approach of the actors is decisive. In this case, individuals are "political actors of a sustainable society made of choices and renunciations" (Buclet 2011, p. 28). They organize themselves to create a network of material flows circulating from one to another and benefit from a reduction in costs.

4.2.4. *The limits of the circular economy*

The circular economy is a radical concept that criticizes the established systemic relationships inherited from contemporary global capitalism. Is it the way out of the dead ends of our business model and towards strong sustainability (Corvellec et al. 2020)? The economy, like environmental innovations, encounters limits related to rebound effects (Bihouix 2014; Korhonen et al. 2018b), the search for eco-efficiency (Arnsperger and Bourg 2016; Korhonen et al. 2018b) and the difficulty of breaking out of the socio-technical regime (Korhonen et al. 2018b; Corvellec et al. 2020).

The limits are also explained by the two principles of thermodynamics. A first category of limitations is related to the first principle, according to which energy and material flows are in a closed system. The circular economy relies on recycling resources, but this is not sufficient for three reasons. First, some materials cannot be recycled due to technical problems (such as thermosetting polymers or soiled medical packaging). Some products cannot be recycled because it is difficult to disassemble them or at least to separate some of their components (such as the metals in a mobile phone). Other components cannot be recovered because of their dispersive use (such as the pigments in paints). The recycling rate is therefore reduced (Bihouix 2014; Arnsperger and Bourg 2016). Second, the beneficial effects of recycling are partial. Indeed, the efficiency of recycling depends on the amount of material it saves (reduced extraction needs). Arnsperger and Bourg (2016) take the example of steel from the work of Grosse. Steel is one of the most recycled major materials in the world, the observed growth rate of its production was 3.25% per year in the 20th century and the current recycling rate is 62%, this finally gaining 12 years against iron scarcity. Samuel Mayer (Box 2.2) also mentions the difficulties in the field of aluminum: "They say that aluminum is infinitely recyclable, but you have to add energy." The beneficial effects of recycling are therefore partial. Moreover, in a growing economy, the long-term benefits of recycling and longer lifetimes are partly offset by the increase in total volumes extracted and consumed. Indeed, due to the growth in consumption in volume, simply because the population is increasing, the resources needed for consumption in t_1 are greater than the resources used in t_0. In other words, the final consumption of resources over time must be less than the available stock. Renewable resources are therefore critical, as a sustainably regenerative economy will be increasingly intensive in exploiting these resources (Arnsperger and Bourg 2016). Third, technology has a critical role, as it enables recycling. The development of industrial ecology, for example, is possible thanks to recycling techniques; it is a primarily technical response (Buclet 2011). These techniques compensate for the inability of the natural system to provide the resources necessary for our mode of production and consumption. The circular economy is therefore part of an approach with low sustainability, where artificial capital (thanks to technology) is substituted for natural capital (natural resources).

Finally, according to the second principle of thermodynamics (Chapter 1), some energy is dissipated during work. Therefore, energy and matter dissipate in circular economy loops (Bihouix 2016; Korhonen et al. 2018b).

In addition to the limitations explained by thermodynamics, three others can be identified. The first is related to the temporal and spatial boundaries of circular economy projects. Most projects are local or regional. Yet a holistic view is needed to measure their contribution to global sustainability (Korhonen et al. 2018b). Similarly, Arnsperger and Bourg (2016) emphasize the need for a macro view to analyze the circular economy. This allows for the identification of rebound effects such as progress on recycling reduced by an increasing consumption of this material. In addition, some end-of-life material flows products cross borders into poor countries, where they degrade the biosphere. A recycling network may reduce its environmental impacts locally in the source country but have negative impacts outside its territory (Korhonen et al. 2018b). Moreover, with the boomerang effect, negative impacts are sent back to the source country. This refers to the decrease in migration of certain species to the source country due to the decrease in biodiversity in the pollution receiving country. A second limitation is related to the ability of circular economy firms to develop inter-organizational and intra-organizational strategies (Korhonen et al. 2018b). Circular economy business models presented earlier show that physical flows of materials and energy cross process, organizational and sector boundaries. New arrangements must be created between producers, suppliers and users to organize the provision of a good. The contracts binding the parties must take into account legal and insurance issues, for example, since the good is no longer sold but loaned. The third limit concerns the definition of physical flows and their assessment in terms of good or bad flows. The definitions are relatively stable with regard to waste management and allow, for example, the building of statistics, as we saw in Chapter 1. However, new operations such as reuse, remanufacturing and renovation are not yet well-defined categories (Korhonen et al. 2018b). Furthermore, the definition of waste depends on cultures, societies and their history. Waste reflects a moral condemnation and is anchored in territorial and cultural dimensions (Corvellec et al. 2020). We have previously indicated that regulatory and culturocognitive institutions explain the change in waste status. Thus, waste did not exist until the end of the 19th century. And today, some forms of waste have become resources and have a market value, which may only be temporary.

4.3. The solutions enabled by sustainable design in the ecological transition perspective

In Chapter 2, we presented sustainable design as the design of new products and new services for the human species while respecting its living environment in a long-term perspective as a key element of its quality of life and the preservation of the planet. Sustainable design is synonymous with environmental and social innovations. We found that innovations with organizational and institutional objectives and based on creative mechanisms had the highest environmental benefits according to the OECD typology. Thus, considering environmental and social impacts together is an approach that can contribute to the ecological transition.

4.3.1. *New possible uses for a new relationship between the human species and its environment*

The pioneers of sustainable design among designers have proposed to design a product by considering the human species in its environment, the object being an intermediary. Petit calls this approach "design of the milieu". Eco-design is an "instrument to compose differently with nature, in an explicit and more careful way" (Bourg and Papaux 2015, p. 313). Antonioli and Vicari (2015, p. 106) present Papanek's approach as a willingness to "look at the real problems of the world and try to solve them by choosing an ecological, rational, and prospective approach at the same time". We can bring this analysis closer to that of Arnsperger and Bourg (2016) on the limits of recycling, which does not allow us in the long-term to solve the equation of resource needs that are growing faster than recycling capacities. Thus, practices that require non-renewable resources and technologies must be limited (Bourg and Papaux 2015, p. 313). Ultimately, only a system based on renewable resources is then feasible. The example of the urban furniture envisioned by the Guyon company (Box 3.2) illustrates a solution based on renewable resources. In other words, the relationship with the environment is modified. We are thus "opposed to an exclusively profit-oriented design practice" (Antonioli and Vicari 2015, p. 106), which proposes goods that are programmed to be obsolete or change according to the diktats of fashion (Petit 2015).

We can thus see the changes in consumption patterns. Kazazian, for his part, observing the inertia of our model at the macro-economic level, claims the essential role of design with companies. The latter constitute the scale to introduce "fundamental changes in the modalities of consumption" with a reform "market tools that are the marketing and the design" (Kazazian 2003, p. 25). Sustainable design can be at the origin of innovations, leading to new behaviors and reducing environmental impact. It can thus lead to changes in usage. According to Petit's (2015) "design of the milieu" approach, the designed object is an intermediary between the individual and the environment. The circular economy business models based on access and performance are then interesting. Indeed, it is no longer the object, but its functionality which becomes commercial. This responds to Kazazian's desire to transform a society of consumption into a society of use with service as its main modality, which in turn leads to a lighter economy (Kazazian 2003).

Furthermore, we observe that certain circular economy business models are the occasion of new relationships between the three spheres (production, use and distribution) in the socio-technical system. The user no longer buys a good from a manufacturer or a distributor (access and performance business model). They are encouraged to buy renovated, repaired or remanufactured goods (business model of the extension of product value), manufactured with recycled materials (business model of the extension of the value of the resource), sustainable (business model of the extension of the life span of the products) or to reduce their consumption (business model of sufficiency). Here, as Papanek wished, the objective of manufacturers and users is to fight against programmed obsolescence and the incessant change of goods due to fashion.

4.3.2. *The advantages of frugal innovation*

We studied frugal innovation in Chapter 2. It can participate in the ecological transition and integrate social issues. One of the characteristics of frugal innovation is to offer products that save resources. This search for eco-efficiency can be explained by the strong constraints on available resources in emerging countries. We have also pointed out that products are more robust, repairable and have a longer life span (Tiwari and Kalogerakis 2020; Levänen et al. 2022).

Frugal innovations are a form of sustainable design associating resource efficiency and accessibility for the poor. Levänen et al. (2022) find from a study of three Indian companies that the technologies used are relatively simple and the resources accessible locally. According to the authors, sustainability is the result of the business model: it makes the product accessible to all and allows for reduced environmental impacts. For example, the company Mitticool offers a clay refrigerator that keeps food or medicine cool. The value proposition is to offer a simple product that works without electricity. The value creation is based on the use of simple technologies, locally available materials and shared tools. Value is captured through the company's ongoing development of products and services, local marketing through word-of-mouth for example and brand name development. Thus, even though imitations of the Mitticool refrigerator are circulating, the enterprise defends its brand by relying on the quality of its product.

Frugal innovation thus contributes to the ecological transition while offering social benefits, as it offers low-cost products thanks to business models adapted to the institutional contexts of developing countries (lack of infrastructure, lack of support from public authorities, for example).

4.4. Perspectives

4.4.1. *Going further by involving stakeholders in innovation*

Some authors emphasize the participatory dimension enabled by sustainable design. According to Antonioli and Vicari (2015), it is about proposing choices to users by exchanging with designers; users contribute to choices that concern their own existence. They are in line with Papanek's approach, for whom technocentric eco-design procedures are not enough and the participation of the users is essential. She thus gives a social ecology dimension to the design. This can be based on the implication of users with participatory or collaborative design (Bourg and Papaux 2015), or on the development of small decentralized production units, with the support of the state, a solution proposed by Papanek (1995, cited by Antonioli and Vicari (2015)).

On a local scale, we are now seeing the development of new design practices that combine ecology (with recycling, repair and reuse) and open participation, in which use value takes precedence over exchange value (Petit 2015). These are fablabs, spaces where individuals can get directly involved in production (Buclet 2014) thanks to digital machine tools, easy to access and use. The value lies in the ability to design and produce on a small scale. Actors and microenterprises share equipment and thus reduce their costs (Bihouix 2014). Buclet (2014) points out that the scale of production promotes exchanges between individuals on their territory. He proposes the example of the Usinette project, which brings together open and shared design objectives and social objectives. It consists of giving each individual the possibility to produce everyday objects. Networks of actors are formed, based on the sharing of knowledge and know-how and on the exchange of materials. Nevertheless, fablabs are not always synonymous with ecology. Bihouix (2014) tempers the enthusiasm around fablabs that depend on 3D printers in particular. Their development is limited, because they are additive technologies (based on polluting materials such as monoplastic polymer) that rather allow monomaterial productions and require raw materials that support fusion.

Within companies, we also observe attempts to open up innovation processes to more actors. Open innovations are approaches that rely on ideas coming from inside and outside the company (Chesbrough 2006). It is the interactions of the company with its institutional, commercial and societal environment that foster innovation (Boutillier et al. 2014). We can mention open labs and living labs. The former are both a place and a process, bringing together several actors. The processes are collaborative, iterative and open, physical or virtual (Le Chaffotec 2016). Living labs, on the contrary, consist of an open innovation process benefiting from the participation of external sources (companies, universities, users), where products and services are tested in real conditions (Le Chaffotec 2016). The customer, user or citizen is solicited "during the research process in conjunction with other stakeholders, companies of all sizes, public institutions public institutions, research laboratories, associations, etc." (Fasshauer and Zadra-Veil 2020, p. 18). Living labs can combine ecological and social objectives. We can cite the example of Nomade Office, a telecommuting and coworking space that is part of sustainable development with "local and bio-sourced materials, consistent with the eco-district of which it is part" (Fasshauer and Zadra-Veil 2020, p. 27).

Approaches combining open innovation and sustainability are visible in companies. Thus, Rauter et al. (2018) carried out a study of 85 Austrian SMEs in industry and services to assess in particular the effects of open innovations on sustainability. They measure sustainable performance according to four criteria: product sustainability, resource efficiency, environmental pollution and social responsibility. Their study identifies positive and significant relationships between sustainable performance and collaboration with different partners. Collaborations with customers (dialogue, analysis of their requirements, participation in innovation) are preferred. To a lesser extent, intermediaries (support in finding partners, source of information on the sector and technologies), suppliers (dialogue on technical developments and activities in joint ventures), associations (innovation platforms and collaborative networks, feedback on innovation projects and assessment of working conditions on production sites), universities (collaboration to improve or develop technologies) and competitors (dialogue on the development of technologies) contribute significantly to sustainable performance.

4.4.2. *Considering the environment as a common good*

The concept of sustainable design combines social and environmental dimensions with a strong focus on sustainability. The work of Elinor Ostrom can provide us with some answers. On the one hand, she is interested in common resources as "a resource system large enough to make it costly (but not impossible) to exclude its potential beneficiaries from access to the benefits of its use" (Ostrom and Baechler 2010, p. 44). Among these resources, the climate must be protected (Coriat 2021). On the other hand, she conducts her analysis with an institutional approach similar to the approach we developed in Chapter 3. The actors are thus the appropriators (who use or consume units of the resources they take), the providers (who organize the supply of resources) and the producers (who undertake activities that build, repair or sustain the resource system).

The environment is a common good, and our society is questioning its management, which is the purpose of the ecological transition. The management system refers to the management of the resource by the actors, the specific rules related to its use and the way these rules are constructed (Ostrom 2009). Ostrom and Baechler (2010) note that most success stories in common pool resource management are associated with institutional arrangements that bring together public and private instruments. Indeed, neither the state nor the market has succeeded in enabling "individuals to make long-term productive use of natural resource systems" (Ostrom and Baechler 2010, p. 14). The solution then lies in governance combining various private actors (the appropriators), public authorities and other civil society actors.

As far as eco-design and its capacity to contribute to the ecological transition are concerned, design approaches associating private and public actors can be a solution. The example of Citiz LPA Grand Lyon is very interesting. It is a semi-public company[8], which offers an electric car sharing service for the Lyon metropolitan area. After the failure of Bluely in 2020, a service that was then managed solely by a private player, Citiz LPA Grand Lyon brings together public players (Grand Lyon, Chamber of Commerce) and a cooperative (Citiz).

4.4.3. *Profound changes are needed in the socio-technical systems*

Are these solutions sufficient? It seems to us that they are not, in the absence of profound changes in the socio-technical systems to support sustainable design approaches. It is worth returning to the elements of the socio-technical systems identified in Chapter 3.

Can companies develop sustainable design methods in relation to users and communities? The functional economy is a business model that requires companies to adapt their offer to better control the life of the product (repair, maintenance). Eco-design then enriches this approach by reducing the environmental impact. Frugal innovations are interesting because they demonstrate the ability to produce with the resources available and accessible to individuals with the lowest incomes. In the long run, this will encourage a change in socio-technical regime supporting the ecological transition.

States can play a major role. Europe enacted the Circular Economy Act in 2015, with eco-design as one of the seven pillars of the circular economy. Promoting sustainable design requires stronger incentives to support participatory open innovation processes. Nevertheless, from the low-carbon electricity transition processes in England and Germany, Geels et al. (2016) demonstrate the difficulty for states to engage in such policies. Governments are under pressure to slow down or even halt policies in favor of decarbonation.

8 Sigot, F. (2021). Lyon Parc Auto multiplie sa flotte d'autopartage par quatre. *Les Échos*, December 23.

Users play an important role in the development of new transition-friendly practices such as open innovation and the functional economy. The sharing economy can also be a solution. It is defined as a business model consisting of sharing, exchanging, renting goods and services without owning them or exchanging goods, services and information between peers (Vernay and Gauthier 2017). These solutions developed by individuals can generate market activities such as carpooling or apartment exchange activities carried out with digital platforms or non-market activities such as local exchange systems where individuals barter services or products. Like the functional economy, where the producer changes its strategy by making a good available, users must change their behavior in favor of using rather than buying goods. These changes in behavior thus lead to a social innovation.

Symbolic meanings guide consumption choices. We have highlighted the long change in social representations regarding the environment. Consumers are generally more in favor of products that are more respectful of the environment. Ecolabels contribute to consumer education. Sustainable design means greater user participation with more collaborative design processes. Examples include consumer cooperatives that manage supermarkets (La Louve in Paris, Demain in Lyon) and cooperatives that group consumers and producers for the production of green energy (Enercoop), which demonstrate the growing interest of individuals in alternative forms of management.

Finally, technical and scientific knowledge in the field of sustainable design must be supported by training in companies and also in initial training. It is a question of increasing the skills in companies, which do not always have this knowledge in-house. This is all the more important as it is necessary to support business models, such as the functional economy, which must also integrate eco-design to reduce pollution. For Karlsson and Luttropp (2006), starting eco-design in companies requires the skills of designers and also the people with whom they must cooperate. It is thus a matter of supporting training programs so that eco-design can spread within the circular economy.

4.5. Conclusion

This chapter focused on the questions and perspectives raised by eco-design in the ecological transition.

The technocentric eco-design method does not make it possible to get out of the technological regime. The technologies are improved and permitted to reduce the environmental impacts in the short-term. They only rarely challenge the organization of production flows: waste and pollution are only limited. It comes up against two obstacles, namely the search for eco-efficiency and the rebound effects. In the long run, it will not solve the equation of the decrease of available resources and pollution.

We have identified the strengths of circular economy business models. The desire to eliminate waste leads to new practices in production and consumption patterns. Among these models, the functional economy (or access and performance model) can contribute to the transition: business strategies and user behaviors are changed in depth. Eco-design practices are then complementary: they evaluate environmental impacts, to propose solutions to reduce them and also to create approaches offering more sustainability, repairability and recyclability.

Nevertheless, the development of the circular economy is confronted with technical limits: recycling does not ultimately allow us to meet the needs of consumption. In other words, technology is not enough. We find the conclusions of the proponents of strong sustainability: capital is not substitutable and natural capital must be preserved. A solution that is consistent with the ecological transition is to develop renewable resources that allow for regenerative systems. It is also necessary to have a macro-economic vision to follow the flows of different resources so that they meet production and consumption needs.

Finally, sustainable design methods constitute interesting alternatives. It is a question of designing with a "design of the milieu" approach, considering the object as an intermediary between the human species and its environment. Thus, designing from renewable resources allows us to break the deadlock based on the extraction of non-renewable resources. Frugal innovation is also promising, because it is based on the principle of access to resources for individuals with the lowest incomes.

We propose two perspectives to develop sustainable design methods: encourage approaches that involve stakeholders in the design process, such as open innovation approaches based on living labs. The common good approach is also interesting. Can we think of a design based on a form of governance that brings together several types of actors: individuals as users, companies, associations and cooperatives, in order to design everyday objects in a sustainable way? This would support a long-term vision that is essential to the ecological transition.

Conclusion

At the end of this book, the assessment is mixed. In the 1960s, enthusiastic designers presented sustainable design as the way out of the dead ends of our market society, where the increasing consumption of objects was synonymous with progress. The chemical industry, on the other hand, was under growing social pressure to limit the pollution it caused. It progressively developed a new design method: technocentric eco-design, based on the idea that it was necessary to evaluate the environmental impacts and then reduce them.

It is clear that more than 50 years after the first texts on sustainable design and technocentric eco-design, a solution allowing the human species to live peacefully in a respected, non-degraded and sustainable environment has not been found.

Technocentric eco-design is the most common and perhaps the easiest to understand and implement. It aims to minimize the environmental impacts of a product compared to similar products on the market and at all stages of its life cycle. It consists of identifying these impacts and then seeking solutions to reduce them. It allows us to improve what already exists, but the pollution remains. Also called environmental design by Petit (2015), it aims to design products by drawing on the resources available to meet the needs of the human species and limiting the pollution emitted. The environment is at the service of the human species. The ecological transition is therefore far away.

For its part, sustainable design seems promising, but it is rarer. We have defined it as the design of new products or services for the human species with respect to its living environment in a long-term perspective as a key element of its quality of life and the preservation of the planet. Described as design of the milieu (Petit 2015), it is a reflection on the object as a mediator between the human species and the environment. Do shared electric vehicles, the New Alchemy Institute's bio-shelters, Guyon's vegetal city furniture and initiatives in fablabs constitute a relevant response? We can sketch out a few avenues for defining a sustainable design method that is likely to meet the challenges of strong sustainability:

– Identify the user's needs in order to provide what is effectively necessary. It is a matter of providing the essential, following the example of frugal innovations. We thus have reflections on low techs (Bihouix 2014), which aim to meet the needs of digital users to provide a service using a minimum of resources.

– Design a product or service in close cooperation with users to find the most relevant ways of using it. This implies closer relationships between actors in a territory. Open innovations and consumer cooperatives can be sources of inspiration.

– Do so only with renewable resources. We have highlighted the limits to recycling non-renewable resources. A circular economy that aims to develop regenerative systems from renewable resources seems the most likely to respect the environment in a sustainable way.

To move towards the ecological transition, certain conditions must be respected:

– The first condition is to develop sustainable infrastructures in the areas of transport, housing, water and energy, for example. Transforming these sectors is difficult. For example, Geels et al. (2016) showed the limits to decarbonization of the electricity sector in England and Germany. Some actors in the system may push or resist change. In Germany, civil society, new green electricity producers and the state have enabled substitution to renewable electricity technologies. In England, the authors find transformation by historical producers, but it is more limited. The dominant technology regime has changed little.

– The second condition requires a strong commitment from the actors. The ecological transition is synonymous with social transformations of lifestyles and technical transformations of production methods. We have mentioned the CSR (corporate social responsibility) strategies of companies that voluntarily support the social or environmental consequences of their activity, but this is not enough. Public policies must provide the framework. They are the manifestation of a collective choice in favor of lifestyles that respect the environment and the well-being of individuals. Nevertheless, it is necessary to take into account the resistance of the actors at the heart of the dominant technological regime, whose positions can be challenged. Civil society therefore has its place. Associations and citizen movements are there to alert and incite change. We mentioned Citizen, a cooperative created by groups of residents in Marseille, Lyon, Grenoble and Strasbourg, and Envie, which brings together the Emmaus association and businesses. Alternatives exist and we must allow them to spread.

The current solutions put forward, such as the circular economy models, consist of preserving the value of products and non-renewable resources as long as possible through reuse, remanufacturing, sustainability and recycling. We have seen that this is not enough to meet the needs of individuals over time.

The environment is a common good and our common concern. To enable current and future generations to live well, it is essential to preserve it, to allow it to last and to reproduce itself today and tomorrow. Olivier Godard (2015, p. 134) indicates that "reproduction does not mean identical perpetuation; it is not a matter of freezing the environment, but of preserving its basic functional organization and its capacity to evolve in a way that preserves its regulations and maintains sufficient diversity of the resources it harbors". It is up to us to change our view of and relationship with the environment.

Between technocentric eco-design and sustainable design, the vision of the environment is not the same. What relationship with the environment do we want to build: a relationship based on harvesting and limiting pollution as much as possible, or a more harmonious relationship, allowing for a sustainable coexistence between the human species and the environment? What means do we give ourselves? It is up to us to write the script for ecological transition.

References

Abrassart, C. (2011). La naissance de l'éco-conception : acteurs, raisonnements, enjeux de pilotage et horizons d'une rationalisation industrielle (1990–2010). PhD Thesis, École nationale supérieure des Mines de Paris, Paris.

Abrassart, C. and Aggeri, F. (2002). La naissance de l'éco-conception. *Responsabilité et environnement*, 25, 14–63.

Abrassart, C. and Aggeri, F. (2008). Le management de l'éco-conception. In *Les nouveaux régimes de la conception : langages, théories, métiers*, Hatchuel, A. and Weil, B. (eds). Vuibert Cerisy, Paris.

ADEME (2009). Les déchets en chiffres en France, Données et références. Document, ADEME, Angers.

ADEME (2014). Coût complet des déchets. Document, ADEME, Angers.

ADEME (2019). La collecte des déchets par le service public en France en 2019 : résultats clés et zooms thématiques. Document, ADEME, Angers.

ADEME (2020a). Guide pratique de la pollution de l'air en 10 questions [Online]. Available at: https://librairie.ademe.fr/cadic/7434/guide-pollution-air-10-questions.pdf.

ADEME (2020b). *Déchets chiffres-clés*, Édition 2020. ADEME, Mulhouse.

ADEME (2022a). Économie circulaire, les impacts environnementaux [Online]. Available at: https://www.ademe.fr/expertises/consommer-autrement/elements-contexte/impacts-environnementaux.

ADEME (2022b). Économie circulaire [Online]. Available at: https://expertises.ademe.fr/expertises/economie-circulaire.

ADEME (2022c). Devenez plus performants grâce à l'écoconception [Online]. Available at: https://expertises.ademe.fr/production-durable/lecoconception/enjeux-lecoconception-benefices-lentreprise-leconomie-lenvironnement.

Aggeri, F., Pezet, E., Abrassart, C., Acquier, A. (2005). *Organiser le développement durable*. Vuibert, Paris.

Aggeri, F., Segrestin, B., Acquier, A. (2006). Régimes de conception et modèles de pilotage de l'éco-conception. Report, ADEME.

Ambec, S. and Lanoie, P. (2008). Does it pay to be green? a systematic overview. *Academy of Management Perspectives*, 22(4), 45–62.

An Inconvenient Truth (2006). Documentary. Directed by Davis Guggenheim. Lawrence Bender Productions and Participant Productions, United States.

Anderson, P. and Tushman, M. (1990). Technological discontinuities and dominant design: A cyclical model of technological change. *Administrative Science Quarterly*, 35(4), 604–633.

Antonioli, M. and Vicari, A. (2015). Archéologie de l'éco-design. In *Poïétiques du design II : Éco-conception ?*, Bertrand, G. and Favard, M. (eds). L'Harmattan, Paris, 101–114.

Arnette, A.N., Brewer, B.L., Choal, T. (2014). Design for sustainability (DFS): The intersection of supply chain and environment. *Journal of Cleaner Production*, 83, 374–390.

Arnsperger, C. and Bourg, D. (2016). Vers une économie authentiquement circulaire. *Revue de l'OFCE*, 145(1), 91–125.

Arundel, A., Kemp, R., Parto, S. (2007). Indicators for environmental innovation: What and how to measure. In *International Handbook on Environment and Technology Management*, Marinova, D., Phillimore, J., Annandale, D. (eds). Edward Elgar, Cheltenham, 324–339.

Asakawa, K., Cuervo-Cazurra, A., Un, C.A. (2019). Frugality-based advantage. *Long Range Planning*, 52(4), 101879.

Aurez, V. and Georgeault, L. (2016a). *Économie circulaire. Système économique et finitude des ressources*. De Boeck Supérieur, Louvain-la-Neuve.

Aurez, V. and Georgeault, L. (2016b). Les indicateurs de l'économie circulaire en Chine. *Revue de l'OFCE*, 1(145), 127–160.

Bahmed, L., Boukhalfa, A., Djebabra, M. (2005). Éco-conception in the industrial firms: Methodological proposition. *Management of Environmental Quality: An International Journal*, 16(5), 530–547.

Bairoch, P. (1997a). *Victoires et déboires. Histoire économique et sociale du monde du XVIe siècle à nos jours*, volume 1. Folio Histoire, Paris.

Bairoch, P. (1997b). *Victoires et déboires. Histoire économique et sociale du monde du XVIe siècle à nos jours*, volume 2. Folio Histoire, Paris.

Bairoch, P. (1997c). *Victoires et déboires. Histoire économique et sociale du monde du XVIe siècle à nos jours*, volume 3, Folio Histoire, Paris.

Bardelli, P. (2016). La flexibilité du travail, un enjeu pour le capitalisme contemporain. *Revue de l'organisation responsable*, 11(1), 5–20.

Bayon, D. and Levratto, N. (2009). *Faire l'économie des déchets : ou quand les dépenses publiques font les profits privés*. Albiana, Ajaccio.

Beaud, M. (2010). *Histoire du capitalisme 1500–2010*, 6th edition. Le Seuil, Paris.

Belin, J., Horbach, J., Oltra V. (2011). Determinants and specificities of eco-innovation, an econometric analysis for the French and German industry based on the community innovation survey. Document, Cahiers du GREThA.

Bellini, B. and Janin, M. (2019). Écoconception : état de l'art des outils disponibles. *Techniques de l'ingénieur*, 4(G6010), G6010–1.

Berneman, C. and Vernier, M.-F. (2011). État de l'art sur la gestion des déchets. Report, ADEME, Angers.

Berneman, C. and Vernier, M.-F. (2014). Dynamiques de la gestion des déchets dans les PME. Report, ADEME, Angers.

Berneman, C., Lanoie, P., Plouffe, S., Vernier, M.-F. (2009). Quels retours économiques pour l'entreprise. Fiches entreprises [Online]. Available at: https://depot.erudit.org/bitstream/003066dd/1/2009s-09.pdf.

Berneman, C., Lanoie, P., Plouffe, S., Vernier, M.-F. (2013). Démystifier la mise en place de l'écoconception. *Gestion*, 38(3), 75–83.

Bihouix, P. (2014). *L'Âge des low tech. Vers une civilisation techniquement soutenable*. Le Seuil, Paris.

Blaug, M. (1985). *La pensée économique. Origine et développement*, 4th edition. Economica, Paris.

Blouet, A. and Rivoire, E. (1995). *L'Écobilan*. Dunod, Paris.

Bocken, N.M. and Short, S.W. (2016). Towards a sufficiency-driven business model: Experiences and opportunities. *Environmental Innovation and Societal Transitions*, 18, 41–61.

Bocken, N.M., De Pauw, I., Bakker, C., Van Der Grinten, B. (2016). Product design and business model strategies for a circular economy. *Journal of Industrial and Production Engineering*, 33(5), 308–320.

Bocquet, Y. (2018). *La mondialisation. Un monde organisé en système*. Éditions universitaires de Dijon, Dijon.

Bonneuil, C. and Frioux, S. (2013). Les "Trente Ravageuses" ? L'impact environnemental et sanitaire des décennies de haute croissance. In *Une autre histoire des "Trente Glorieuses". Modernisation, contestations et pollutions dans la France d'après-guerre*, Pessis, C., Topçu, S., Bonneuil, C. (eds). La Découverte, Paris.

Bourg, D. and Papaux, A. (2015). *Dictionnaire de la pensée écologique*. PUF, Paris.

Bourg, D. and Rayssac, A. (2006). *Le développement durable : maintenant ou jamais*. Gallimard, Paris.

Boutaud, A. and Gondran, N. (2018). *L'empreinte écologique*. La Découverte, Paris.

Boutillier, S. (2003). Les économistes et l'écologie, enseignements historiques. *Innovations*, 18(2), 139–165.

Boutillier, S., Djellal, F., Gallouj, F., Laperche, B., Uzunidis, D. (2012). *L'innovation verte. De la théorie aux bonnes pratiques*. Peter Lang, Brussels.

Boutillier, S., Forest, J., Gallaud, D., Laperche, B., Tanguy, C., Temri, L. (2014). *Principes d'économie de l'innovation*. Peter Lang, Brussels.

Braungart, M., McDonough, W., Bollinger, A. (2007). Cradle-to-cradle design: Creating healthy emissions – A strategy for eco-effective product and system design. *Journal of Cleaner Production*, 15(13/14), 1337–1348.

Brodhag, C. (2014). L'écoconception, un outil au cœur de l'économie circulaire. *Annales des Mines – Responsabilité et environnement*, 4(76), 33–37.

Bruel, A., Kronenberg, J., Troussier, N., Guillaume, B. (2019). Linking industrial ecology and ecological economics: A theoretical and empirical foundation for the circular economy. *Journal of Industrial Ecology*, 23(1), 12–21.

Brundtland, G.H. (1987). Our common future. From one earth to one world. *World Commission on Environment and Development* [Online]. Available at: https://sustainabledevelopment.un.org/content/documents/5987our-common-future.pdf.

Buckminster Fuller, R. and McHale, J. (1963). *World Science Decade 1965–1975, Phase 1. Document 1: Inventory of World Resources Human Trends and Needs*. World Resources Inventory, Southern Illinois University, Carbondale.

Buckminster Fuller, R. and McHale, J. (1964). *World Science Decade 1965–1975, Phase 1. Document 2: The Design Initiative*. World Resources Inventory, Southern Illinois University, Carbondale.

Buclet, N. (2011). *Écologie industrielle et territoriale : stratégies locales pour un développement durable*. Presses Universitaires du Septentrion, Villeneuve-d'Ascq.

Buclet, N. (2014). L'économie de fonctionnalité entre éco-conception et territoire : une typologie. Développement durable et territoires. *Développement durable & territoires. Économie, géographie, politique, droit, sociologie*, 5(1), 1–17.

Bürgenmeier, B. (2008). *Politiques économiques du développement durable*. De Boeck Supérieur, Louvain-la-Neuve.

Cainelli, G., D'Amato, A., Mazzanti, M. (2020). Resource efficient eco-innovations for a circular economy: Evidence from EU firms. *Research Policy*, 49(1), 103827.

Capron, M. and Quairel-Lanoizelée, F. (2016). *La responsabilité sociale d'entreprise*. La Découverte, Paris.

Carayannis, E.G. and Campbell, D.F.J. (2017). Les systèmes d'innovation de la quadruple et de la quintuple hélice. *Innovations*, 3(54), 173–195.

Chesbrough, H. (2006). Open innovation: A new paradigm for understanding industrial innovation. In *Open Innovation: Researching a New Paradigm*, Chesbrough, H., Vanhaverbeke, W., West, J. (eds). Oxford University Press, New York.

Chessel, M.-E. (2012). Introduction : Les constructeurs de la société de consommation. In *Histoire de la consommation*, Chessel, M.-E. (ed.). La Découverte, Paris.

Chia, R. and Nayak, A. (2012). Décisions dramatiques ou incisions imperceptibles ? *Revue française de gestion*, 225(6), 147–166.

Clément, M., Douai, A., Gondard-Delcroix, C. (2011). Réflexions sur le concept de soutenabilité sociale dans le contexte des pays du Sud. *Mondes en développement*, 4(156), 7–18 [Online]. Available at: https://doi.org/10.3917/med.156.0007.

Cloutier, J. (2003). Qu'est-ce que l'innovation sociale ? *Les cahiers du CRISES*, ET0314, 1–46.

Coenen, L. and López, F.J.D. (2010). Comparing systems approaches to innovation and technological change for sustainable and competitive economies: An explorative study into conceptual commonalities, differences and complementarities. *Journal of Cleaner Production*, 18(12), 1149–1160.

Commission européenne (2001a). Livre vert, du 7 février 2001, sur la politique intégrée des produits [Online]. Available at: https://eur-lex.europa.eu/FR/legal-content/summary/integrated-product-policy.html.

Commission européenne (2001b). Livre vert, promouvoir un cadre européen pour la responsabilité sociale des entreprises. July 18 [Online]. Available at: https://eur-lex.europa.eu/legal-content/FR/ALL/?uri=celex:52001DC0366.

Commission européenne (2005). Directives sur les déchets des équipements électriques et électroniques [Online]. Available at: https://ec.europa.eu/environment/waste/weee/old_en.htm.

Commission européenne (2007). Directives REACH et ROHS [Online]. Available at: https://ec.europa.eu/environment/waste/rohs_eee/legis_rohs1_en.htm.

Commission européenne (2008). Directive 2008/98/ce du parlement européen et du conseil relative aux déchets et abrogeant certaines directives [Online]. Available at: https://eur-lex.europa.eu/legal-content/FR/TXT/HTML/?uri=CELEX:32008L0098&from=EN.

Commission européenne (2009). Directive éco-design [Online]. Available at: https://ec.europa.eu/info/energy-climate-change-environment/standards-tools-and-labels/products-labelling-rules-and-requirements/energy-label-and-ecodesign/energy-efficient-products_fr.

Commission européenne (2016). European SMEs and the circular economy [Online]. Available at: https://ec.europa.eu/environment/green-growth/docs/fl_441_sum_en.pdf.

Commission européenne (2020). Key consumer data [Online]. Available at: https://ec.europa.eu/info/policies/consumers/consumer-protection-policy/key-consumer-data_en [Accessed 23 August 2022].

Coppin, O. (2002). Le milieu innovateur : une approche par le système. *Innovations*, 2(16), 29–50.

Coriat, B. (2021). *Le bien commun, le climat et le marché : réponse à Jean Tirole*. Les Liens qui libèrent, Paris.

Corteel, D. and Le Lay, S. (2011). *Les travailleurs des déchets*. Érès, Toulouse.

Corvellec, H., Böhm, S., Stowell, A., Valenzuela, F. (2020). Introduction to the special issue on the contested realities of the circular economy. *Culture and Organization*, 26(2), 97–102.

CRÉDOC (2018). Consommer durable est-il un acte de distinction ? Représentations, pratiques et impacts écologiques réels au regard des dynamiques sociales. CRÉDOC, 344, 1–175.

CRÉDOC (2019). Consommation durable : l'engagement de façade des classes supérieures. CRÉDOC, 303, March.

DaSilva, C.M. and Trkman, P. (2014). Business model: What it is and what it is not. *Long Range Planning*, 47(6), 379–389.

Daumas, L. (2020). L'effet-rebond condamne-t-il la transition à l'échec ? *Regards croisés sur l'économie*, 26(1), 189–197.

De Jesus, A. and Mendonça, S. (2018). Lost in transition? Drivers and barriers in the eco-innovation road to the circular economy. *Ecological Economics*, 145, 75–89.

De Perthuis, C. (2014). Économie circulaire et transition écologique. *Annales des Mines – Responsabilité et environnement*, 4(76), 23–27.

Debref, R. (2016). Pour une approche systémique de l'innovation "environnementale". *Revue d'économie industrielle*, 155, 71–98.

Debref, R. (2017). Revising boundaries of the process of environmental innovation to prevent climate change. *Journal of Innovation Economics Management*, 24(3), 9–34.

Debref, R. (2018). *Environmental Innovation and Ecodesign: Certainties and Controversies*. ISTE Ltd, London, and John Wiley & Sons, New York.

Dequech, D. (2003). Cognitive and cultural embeddedness: Combining institutional economics and economic sociology. *Journal of Economic Issues*, 37(2), 461–470.

Designers + (2014). Eco-Design ? Rôle, processus et outils pour le designer [Online]. Available at: https://www.eco-conception.fr/library/h/eco-design--role-processus-et-outils-pour-le-designer.html.

Díaz-García, C., González-Moreno, Á., Sáez-Martínez, F.J. (2015). Eco-innovation: Insights from a literature review. *Innovation: Organization & Management*, 17(1), 6–23.

Diemer, A. (2016). Les symbioses industrielles : un nouveau champ d'analyse pour l'économie industrielle. *Innovations*, 2(50), 65–94.

Elkington, J. (2013). Enter the triple bottom line. In *The Triple Bottom Line, Does It All Add up? Assessing the Sustainability of Business and CSR*, Henriques, A. and Richardson, J. (eds). Routledge, London.

Fasshauer, I. and Zadra-Veil, C. (2020). Le living lab, un intermédiaire d'innovation ouverte pour les territoires ruraux ou péri-urbains ? *Innovations*, 1(61), 15–40.

Fava, J.A. (2005). Can ISO life cycle assessment standards provide credibility for LCA? *Building Design & Construction*, 17–20.

Figuière, C., Boidin, B., Diemer, A. (2014). *Économie politique du développement durable*, 1st edition. De Boeck Supérieur, Louvain-la-Neuve.

Fontan, J.M. (2008). Développement territorial et innovation sociale : l'apport polanyien. *Revue Interventions économiques. Papers in Political Economy*, 38, 1–18.

Forti, V., Balde, C.P., Kuehr, R., Bel, G. (2020). The global E-waste Monitor, 2020: Quantities, flows and the circular economy potential. Report, United Nations University (UNU)/United Nations Institute for Training and Research (UNITAR)/SCYCLE Programme/International Telecommunication Union (ITU) & International Solid Waste Association (ISWA), Bonn/Geneva/Rotterdam [Online]. Available at: https://ewaste monitor.info/gem-2020/.

Fournier, M., Grison, J.B., Rieutort, L. (2014). D'une innovation sociale à la définition d'un projet de territoire : la capacité mobilisatrice d'une association d'insertion dans un milieu montagnard marginalisé. L'exemple des Ateliers de la Bruyère au Pays de Saugues. In *La transformation sociale par l'innovation sociale, 4e Colloque international du laboratoire CRISES*, Université du Québec à Montréal (UQAM), 3–4 April, Montreal.

Frioux, S. (2013). La pollution de l'air, un mal nécessaire ? La gestion du problème durant les "trente pollueuses". In *Une autre histoire des "Trente Glorieuses". Modernisation, contestations et pollutions dans la France d'après-guerre*, Pessis, C., Topçu, S., Bonneuil, C. (eds). La Découverte, Paris.

Frishammar, J. and Parida, V. (2019). Circular business model transformation: A roadmap for incumbent firms. *California Management Review*, 61(2), 5–29.

Froger, G., Calvo-Mendieta, I., Petit, O., Vivien, F.D. (2016). Qu'est-ce que l'économie écologique ? *L'Économie politique*, 69(1), 8–23.

Frosch, R.A. and Gallopoulos, N.E. (1989). Strategies for manufacturing. *Scientific American*, 261(3), 144–153.

Gallaud, D. and Laperche, B. (2016). *Circular Economy, Industrial Ecology and Short Supply Chain*. ISTE Ltd, London, and John Wiley & Sons, New York.

Galliano, D. and Nadel, S. (2013). Les déterminants de l'adoption de l'écoinnovation selon le profil stratégique de la firme : le cas des firmes industrielles françaises. *Revue d'économie industrielle*, 2(142), 77–110.

Garrido-Prada, P., Lenihan, H., Doran, J., Rammer, C., Perez-Alaniz, M. (2021). Driving the circular economy through public environmental and energy R&D: Evidence from SMEs in the European Union. *Ecological Economics*, 182, 106884.

Gasmi, N. and Grolleau, G. (2003). Spécificités des innovations environnementales. *Innovations*, 2(18), 73–89.

Geels, F.W. (2004). From sectoral systems of innovation to socio-technical systems: Insights about dynamics and change from sociology and institutional theory. *Research Policy*, 33(6/7), 897–920.

Geels, F.W. (2014). Regime resistance against low-carbon transitions: Introducing politics and power into the multi-level perspective. *Theory, Culture & Society*, 31(5), 21–40.

Geels, F.W. and Schot, J. (2007). Typology of sociotechnical transition pathways. *Research Policy*, 36(3), 399–417.

Geissdoerfer, M., Savaget, P., Bocken, N.M., Hultink, E.J. (2017). The circular economy – A new sustainability paradigm? *Journal of Cleaner Production*, 143, 757–768.

Ghisetti, C. and Montresor, S. (2020). On the adoption of circular economy practices by small and medium-size enterprises (SMEs): Does "financing-as-usual" still matter? *Journal of Evolutionary Economics*, 30(2), 559–586.

Godard, O. (2015). *Environnement et développement durable : une approche méta-économique*. De Boeck Supérieur, Louvain-la-Neuve.

Goodwin, N., Nelson, J.A., Ackerman, F., Weisskopf, T. (2008). *Consumption and the Consumer Society*. Global Development and Environment Institute (GDAE), Tufts University, Medford.

GreenFlex (2021). Baromètre 2021 de la consommation responsable [Online]. Available at: https://info.greenflex.com/contenu-barom%C3%A8tre-greenflex-ademe-2021-focus-sectoriel.

Greeno, J. and Robinson, S. (1992). Rethinking corporate environmental management. *The Columbia Journal of World Business*, 27(3/4), 222–232.

Grison, J.-B., Rieutort, L., Fournier, M. (2018). Quelle innovation pour un territoire et une ressource marginalisés ? L'exemple de la filière laine en Margeride. In *Les campagnes européennes : espaces d'innovations dans un monde urbain*, Margetic, C., Roth, H., Pouzenc, M. (eds). Presses universitaires du Midi, Toulouse.

Groff, J. and Nguyen-Thi, T.U. (2012). Motivations à l'éco-innovation : une comparaison sectorielle sur les entreprises au Luxembourg. Working document, CEPS Instead, 11.

Guilloux, G. (2015). Du contexte au produit : intégration de l'environnement aux métiers du design. In *Poïétiques du design II : Eco-conception ?* Bertrand, G. and Favard, M. (eds). L'Harmattan, Paris.

Hatchuel, A. (2006). Quelle analytique de la conception ? Parure et pointe en design. In *Le design. Essais sur des théories et des pratiques*, Flamand, B. (ed.). Éditions du Regard, Paris.

Haurand, M.D. and Stummer, C. (2018). Stakes or garlic? Studying the emergence of dominant designs through an agent-based model of a vampire economy. *Central European Journal of Operations Research*, 26(2), 373–394.

Hekkert, M.P., Suurs, R.A., Negro, S.O., Kuhlmann, S., Smits, R.E. (2007). Functions of innovation systems: A new approach for analysing technological change. *Technological Forecasting and Social Change*, 74(4), 413–432.

Hillier, J., Moulaert, F., Nussbaumer, J. (2004). Trois essais sur le rôle de l'innovation sociale dans le développement spatial. *Géographie, économie société*, 2(6), 129–152.

den Hollander, M.C., Bakker, C.A., Hultink, E.J. (2017). Product design in a circular economy: Development of a typology of key concepts and terms. *Journal of Industrial Ecology*, 21(3), 517–525.

Horbach, J. (2008). Determinants of environmental innovation: New evidence from German panel data sources. *Research Policy*, 37(1), 163–173.

Horbach, J., Rammer, C., Rennings, K. (2012). Determinants of eco-innovations by type of environmental impact – The role of regulatory push/pull, technology push and market pull. *Ecological Economics*, 78, 112–122.

INSEE (2022). Innovation [Online]. Available at: https://www.insee.fr/fr/metadonnees/definition/c1182.

IPCC (2021). Summary for policymakers. In *Climate Change 2021: The Physical Science Basis. Contribution of Working Group I to the Sixth Assessment Report of the Intergovernmental Panel on Climate Change*, Masson-Delmotte, V., Zhai, P., Pirani, A., Connors, S.L., Péan, C., Berger, S., Caud, N., Chen, Y., Goldfarb, L., Gomis, M.I. et al. (eds). Cambridge University Press, Cambridge/New York.

Janin, M. (2000). Démarche d'éco-conception en entreprise : un enjeu, construire la cohérence entre outils et processus. PhD Thesis, ENSAM, Paris.

Johansson, G., Widheden, J., Bergendahl, C.G. (2001). Green is the colour of money-commercial success stories from ecodesign. Report, GreenPack, Nordic Industrial Fund.

Karlsson, R. and Luttropp, C. (2006). EcoDesign: What's happening? An overview of the subject area of EcoDesign and of the papers in this special issue. *Journal of Cleaner Production*, 14(15/16), 1291–1298.

Kazazian, T. (2003). *Il y aura l'âge des choses légères*. Victoires Éditions, Paris.

Kirchherr, J., Piscicelli, L., Bour, R., Kostense-Smit, E., Muller, J., Huibrechtse-Truijens, A., Hekkert, M. (2018). Barriers to the circular economy: Evidence from the European Union (EU). *Ecological Economics*, 150, 264–272.

Knight, P. and Jenkins, J.O. (2009). Adopting and applying eco-design techniques: A practitioners perspective. *Journal of Cleaner Production*, 17(5), 549–558.

Korhonen, J., Nuur, C., Feldmann, A., Birkie, S.E. (2018a). Circular economy as an essentially contested concept. *Journal of Cleaner Production*, 175, 544–552.

Korhonen, J., Honkasalo, A., Seppälä, J. (2018b). Circular economy: The concept and its limitations. *Ecological Economics*, 143, 37–46.

Lanoie, P. and Plouffe, S. (2008). Quels retours économiques pour l'entreprise. État de l'art. Pôle national Éco-conception, Saint-Étienne, Institut de Développement Produit, Montreal.

Laperche, B. and Merlin-Brogniart, C. (2020). Les systèmes produit-service (SPS) : leviers de la transition des économies industrielles. *Technologie et innovation*, 5(1), 1–11.

Laperche, B. and Uzunidis, D. (2007). Le Système national d'innovation russe en restructuration. *Innovations*, 2(26), 69–94.

Le Bas, C. (2017). Sustainable innovation and frugal innovation: Exploring the relationships between innovation and sustainability. *Revue d'économie industrielle*, 3(159), 113–137.

Le Bas, C. and Khan, S. (2021). Frugality – Frugal innovation as inclusive innovation. In *Innovation Economics, Engineering and Management Handbook 1: Main Themes*, Uzunidis, D., Kasmi, F., Adatto, L. (eds). ISTE Ltd, London, and John Wiley & Sons, New York.

Le Chaffotec, A. (2016). Quoi de neuf, Docteur ? Les Living Labs en santé. *Entreprendre & Innover*, 4(31), 25–35.

Le Moigne, R. (2014). *L'économie circulaire. Comment la mettre en œuvre dans l'entreprise grâce à la reverse supply-chain*. Dunod, Paris.

Le Pochat, S. (2005). Intégration de l'éco-conception dans les PME : proposition d'une méthode d'appropriation de savoir-faire pour la conception environnementale des produits. PhD Thesis, ENSAM, Paris.

Levänen, J., Hossain, M., Wierenga, M. (2022). Frugal innovation in the midst of societal and operational pressures. *Journal of Cleaner Production*, 347, 131308.

Lifset, R. and Graedel, T.E. (2002). Industrial ecology: Goals and definitions. In *A Handbook of Industrial Ecology*, Ayres, R.U. and Ayres, L.W. (eds). Edward Elgar Publishing Ltd., Cheltenham.

Lopez, F.J.D., Bastein, T., Tukker, A. (2019). Business model innovation for resource-efficiency, circularity and cleaner production: What 143 cases tell us. *Ecological Economics*, 155, 20–35.

Lüdeke-Freund, F., Gold, S., Bocken, N.M. (2019). A review and typology of circular economy business model patterns. *Journal of Industrial Ecology*, 23(1), 36–61.

Lundvall, B.Å. (2016). *The Learning Economy and the Economics of Hope*. Anthem Press, London and New York.

Lupton, S. (2011). *Économie des déchets. Une approche institutionnaliste*. De Boeck, Brussels.

Maccioni, L., Borgianni, Y., Pigosso, D.C.A. (2019). Can the choice of eco-design principles affect products' success? *Design Science*, 5, 1–31.

Madge, P. (1997). Ecological design: A new critique. *Design Issues*, 13(2), 44–54.

Manzini, E. (1991a). The limits and possibilities of eco-design. *Workshop d'experts "Ecodesign of Products"*, Delft.

Manzini, E. (1991b). *Artefacts. Vers une nouvelle écologie de l'environnement artificiel*. Éditions du Centre Pompidou, Paris.

Manzini, E. and Vezzoli, C. (2003). A strategic design approach to develop sustainable product service systems: Examples taken from the "environmentally friendly innovation" Italian prize. *Journal of Cleaner Production*, 11(8), 851–857.

Margni, M., Munch, F., Plouffe, S., Sauvé, S., Pelletier, J., Rebolledo, C., Beaulieu, M., Roy, J., Ponce, S. (2016). Les outils, les ressources et les savoirs existants. In *L'économie circulaire : une transition incontournable*, McDonald, M., Normandin, D., Sauvé, S. (eds). Presses de l'Université de Montréal, Montreal.

Masoumeh, H.S. and Maryam, N.S. (2020). Business model innovation as a turnaround strategy. *Journal of Strategy and Management*, 13(2), 241–253.

Mendoza, J.M.F., Sharmina, M., Gallego-Schmid, A., Heyes, G., Azapagic, A. (2017). Integrating backcasting and eco-design for the circular economy: The BECE framework. *Journal of Industrial Ecology*, 21(3), 526–544.

Meyer, J.W. and Rowan, B. (1991). Institutionalized organizations: Formal structure as myth and ceremony. In *The New Institutionalism in Organizational Analysis*, Powel, W.W. and Di Maggio, P.J. (eds). The University of Chicago Press, Chicago.

Millet, D. (1995). Prise en compte de l'environnement en conception : proposition d'une démarche d'aide à la conception permettant de limiter les ponctions et rejets engendrés par le produit sur son cycle de vie. PhD Thesis, ENSAM, Paris.

Monnoyer-Smith, L. (2017). Transition numérique et transition écologique. *Annales des Mines – Responsabilité et environnement*, 3(87), 5–7.

Nelson, R.R. and Winter, S.G. (1982). *An Evolutionary Theory of Economic Change*. Harvard University Press, Cambridge.

North, D.C. (1991). Institutions. *Journal of Economic Perspectives*, 5(1), 97–112.

O'Brien, M., Miedzinski, M., Giljum, S., Doranova, A. (2014). Eco-innovation and competitiveness: Enabling the transition to a resource-efficient circular economy. Report, European Commission.

OECD (2009). Sustainable manufacturing and eco-innovation: Towards a green economy. *OECD Observer*, June.

OECD and Eurostat (2005). *Manuel d'Oslo : principes directeurs pour le recueil et l'interprétation des données sur l'innovation. La mesure des activités scientifiques et technologiques*, 3rd edition. OECD, Paris [Online]. Available at: https://doi.org/10.1787/9789264013124-fr.

Ostrom, E. (2009). A general framework for analyzing sustainability of social-ecological systems. *Science*, 325(5939), 419–422.

Ostrom, E. and Baechler, L. (2010). *Gouvernance des biens communs*. De Boeck, Brussels.

Papanek, V. (1974). *Design pour un monde réel : écologie humaine et changement social*. Mercure de France, Paris.

Papanek, V. (1995). *The Green Imperative: Ecology and Ethics in Design and Architecture*. Thames and Hudson Ltd, London.

Pasquale, D.V., Secundo, G., Mele, G., Passiante, G. (2021). Sustainable entrepreneurship education for circular economy: Emerging perspectives in Europe. *International Journal of Entrepreneurial Behaviour & Research*, 27(8), 2096–2124.

Patingre, J.F. (2001). L'analyse fonctionnelle des produits en éco-conception, In *Éco-conception : concept, méthodes, outils, guides et perspectives*, Vigneron, J. and Patingre, J.F. (eds). Economica, Paris.

Pereira, Á. and Vence, X. (2020). Les systèmes produit-service comme business models pour l'économie circulaire : potentialités et limites. *Technologie et innovation*, 5(1), 1–22.

Pessis, C., Topçu, S., Bonneuil, C. (2013). *Une autre histoire des "Trente Glorieuses". Modernisation, contestations et pollutions dans la France d'après-guerre*. La Découverte, Paris.

Petit, V. (2015). L'éco-design : design de l'environnement ou design du milieu ? *Sciences du design*, 2(2), 31–39.

Pieroni, M.P.P., McAloone, T.C., Pigosso, D.C.A. (2019). Business model innovation for circular economy and sustainability: A review of approaches. *Journal of Cleaner Production*, 215, 198–216.

Plihon, D. (2007). Capitalisme. In *Dictionnaire de l'économie, encyclopédie Universalis*. Albin Michel, Paris.

Plouffe, S., Lanoie, P., Berneman, C., Vernier, M.-F. (2011). Economic benefits tied to ecodesign. *Journal of Cleaner Production*, 19(6/7), 573–579.

Polanyi, K. (1983). *La grande transformation*. Gallimard, Paris.

Pôle éco-conception (2007). Traité d'éco-conception. Manual, Pôle éco-conception, Saint-Étienne.

Pôle éco-conception (2016). Fiche Éco-initiative 10. Pôle éco-conception, Saint-Étienne [Online]. Available at: https://www.eco-conception.fr/library/h/fiche-eco-initiative-seb-eurecook.html.

Pôle éco-conception (2017). Guide outil d'éco-conception. Pôle éco-conception, Saint-Étienne.

Pôle éco-conception (2018). Fiche Éco-initiative 12. Pôle éco-conception, Saint-Étienne [Online]. Available at: https://www.eco-conception.fr/library/h/fiche-eco-initiative-epta-bonnet-neve.html.

Pôle éco-conception (2021). Fiche Éco-initiative 23. Pôle éco-conception, Saint-Étienne [Online]. Available at: https://www.eco-conception.fr/library/h/eco-initiative-alstom.html.

Pôle éco-conception (2022a). Histoire de l'ACV. Pôle éco-conception, Saint-Étienne [Online]. Available at: https://www.eco-conception.fr/static/histoire-acv.html.

Pôle éco-conception (2022b). L'éco-conception. Les concepts [Online]. Available at: https://www.eco-conception.fr/static/leco-conception-les-concepts.html.

Postel, N. and Rousseau, S. (2008). RSE et éthique d'entreprise : la nécessité des institutions. *M@n@gement*, 11(2), 137–160.

Quairel-Lanoizelée, F. (2011). La responsabilité sociale des entreprises est-elle soluble dans la concurrence ? In *La responsabilité sociale de l'entreprise : nouvelle régulation du capitalisme ?*, Postel, N., Chavy, F., Sobel, R., Doucin, M. (eds). Presses Universitaire du Septentrion, Villeneuve-d'Ascq.

Ratzel, F. (1891). *Anthropogéographie*. J. Engelhorn, Stuttgart.

Rauter, R., Globocnik, D., Perl-Vorbach, E., Baumgartner, R.J. (2019). Open innovation and its effects on economic and sustainability innovation performance. *Journal of Innovation & Knowledge*, 4(4), 226–233.

Rennings, K. (2000). Redefining innovation – Eco-innovation research and the contribution from ecological economics. *Ecological Economics*, 32(2), 319–332.

Rennings, K. and Rammer, C. (2009). Increasing energy and resource efficiency through innovation: An explorative analysis using innovation survey data. *Czech Journal of Economics and Finance*, 59(5), 442–459.

Rexhäuser, S. and Rammer, C. (2014). Environmental innovations and firm profitability: Unmasking the Porter hypothesis. *Environmental and Resource Economics*, 57(1), 145–167.

Ricœur, P. (1990). *Soi-même comme un autre*. Le Seuil, Paris.

Roux, P., Patingre, J.F., Giroux, F., Naud, O. (2005). Éco-technologies, écoconception, évaluation intégrée. Session, Académie d'Agriculture de France, May 18.

Saidani, M., Yannou, B., Leroy, Y., Cluzel, F., Kendall, A. (2019). A taxonomy of circular economy indicators. *Journal of Cleaner Production*, 207, 542–559.

Schumpeter, J. (1981). *Theory of Economic Development*. Transaction Publishers, Piscataway.

Scott, W.R. (2008). *Institutions and Organizations. Ideas and Interests*. Sage, Thousand Oaks.

Serveille H. and Lavergne, R. (2017). Introduction. *Annales des Mines – Responsabilité et environnement*, 3(87), 8–10.

SETAC (1991). A technical framework for life-cycle assessment. Report, Society of Environmental Toxicology and Chemistry (SETAC), Washington.

Silpa, K., Yao, L., Bhada-Tata, P., Van Woerden, F. (2018). *What a Waste 2.0: A Global Snapshot of Solid Waste Management to 2050*. World Bank Group, Washington. doi:10.1596/978-1-4648-1329-0.

Spangenberg, J.H., Fuad-Luke, A., Blincoe, K. (2010). Design for Sustainability (DfS): The interface of sustainable production and consumption. *Journal of Cleaner Production*, 18(15), 1485–1493.

Stamm C.B. (2015). Si la transition écologique avait lieu... Une prospective sociologique pour élargir la discussion sur la responsabilité des entreprises. *Revue de l'organisation responsable*, 10(2), 75–87.

Steelcase (2022). Do Chairs. *Environmental Product Declaration* [Online]. Available at: https://www.environdec.com/library/epd2981.

Suchman, C.M. (1995). Managing legitimacy: Strategic and institutional approaches. *Academy of Management Review*, 20(3), 571–610.

Teece, D.J. (2010). Business models, Business strategy and innovation. *Long Range Planning*, 43(2/3), 172–194.

The Day After Tomorrow (2004). Film. Directed by Roland Emmerich. Centropolis Entertainment, Lionsgate Films, The Mark Gordon Company, United States.

Tiwari, R. and Herstatt, C. (2020). "Pushing the envelope" – Transcending the conventional wisdom of frugal innovation. *International Journal of Technology Management*, 83(1/2/3), 1–14.

Tiwari, R. and Kalogerakis, K. (2020). What enables frugal innovation? An examination of innovation pathways in India's auto component industry. In *Managing Innovation in a Global and Digital World*, Tiwari, R. and Buse, S. (eds). Springer Gabler, Wiesbaden.

Todd, N.J. (2005). *Ecodesign : des solutions pour la planète. L'aventure du New Alchemy Institute*. Écosociété, Montreal.

Tran Dieu, L. and Vernier, M.-F. (2017). La prévention des déchets : une analyse empirique des déterminants du comportement des entreprises. *Revue d'économie industrielle*, 3(159), 79–111.

Tremblay, D.G. (2014). *L'innovation technologique, organisationnelle et sociale*. Presses de l'Université du Québec, Quebec.

Trischler, J., Johnson, M., Kristensson, P. (2020). A service ecosystem perspective on the diffusion of sustainability-oriented user innovations. *Journal of Business Research*, 116, 552–560.

Tukker, A., Haag, E., Eder, P., Vercalsteren, A., Tischner, T.W.U., Charter, M., van der Vlugt, M. (2000). Eco-design: European state of the art. Report, ESTO, Institute for Prospective Technological Studies, Seville.

Urbinati, A., Chiaroni, D., Chiesa, V. (2017). Towards a new taxonomy of circular economy business models. *Journal of Cleaner Production*, 168, 487–498.

Urbinati, A., Franzò, S., Chiaroni, D. (2021). Enablers and barriers for circular business models: An empirical analysis in the Italian automotive industry. *Sustainable Production and Consumption*, 27, 551–566.

Vaileanu-Paun, I. and Boutillier, S. (2012a). Développement durable. Du paradigme de l'économie industrielle à l'économie de la fonctionnalité. In *L'innovation verte – de la théorie aux bonnes pratiques*, Boutillier, S., Djellal, F., Gallouj, F., Laperche, B., Uzunidis, D. (eds). Peter Lang, Brussels.

Vaileanu-Paun, I. and Boutillier, S. (2012b). Économie de la fonctionnalité. Une nouvelle synergie entre le territoire, la firme et le consommateur ? *Innovations*, 1(37), 95–125.

Vence, X. and Pereira, Á. (2019). Eco-innovation and circular business models as drivers for a circular economy. *Contaduría y Administración*, 64(1), 1–19.

Vernay, A. and Gauthier, C. (2017). Tempête dans le secteur de l'électricité : l'émergence de l'économie du partage aux Pays-Bas. *Entreprises et histoire*, 1(86), 105–124.

Vernier, M.-F. (2010a). Éthique sur le marché : comment permettre le succès de produit intégrant une démarche éthique. Le cas de l'éco-conception. *Éthique et économique/Ethics and economics*, 7(1), 1–22.

Vernier, M.-F. (2010b). Fiche 21. Les exemples de Lafuma et de Steelcase, In *50 fiches sur le marketing durable*, Lavorata, L. (ed.). Bréal, Paris.

Vernier, M.-F. (2013). Les facteurs clés de succès de l'écoconception : leçons tirées de quelques cas d'entreprises. *Gestion*, 38(3), 67–74.

Vernier, M.-F. (2023). De l'éco-conception à l'innovation environnementale. Chronique d'un échec. In *Imaginer le futur à partir du passé*, Forest J. and Bocquet, B. (eds). Peter Lang, Bern.

Vernier, M.-F., Arfaoui, N., Challita, S., Lanoie, P., Plouffe, S. (2022). Factors influencing profitability in eco-design: Lessons from European and Canadian firms. *Journal of Innovation Economics & Management*, 3(39), 141–172.

Veyret, Y. (2016). *Le développement durable*. UPPR Éditions, Toulouse.

Vidalenc, E., Bergey, J.-L., Quiniou, V., Marchal, D., Combet, E. (2022). Quatre scénarios pour la transition écologique. L'exercice de prospective de l'ADEME Transition(s) 2050. *Futuribles*, 3(448), 5–21.

Vigneron, J. and Patingre, J.F. (2001). *Éco-conception : concept, méthodes, outils, guides et perspectives*. Economica, Paris.

Vincent-Sweet, P. (2012). Analyse du cycle de vie et protection de l'environnement : pertinence et limites de l'outil. Le point de vue d'une association. *Annales des Mines – Responsabilité et environnement*, 2(66), 84–88.

Vivien, F.D. (2003). Rencontre du troisième type... d'écosystème ou quand l'écologie devient industrielle. *Innovations*, 2(18), 43–57.

Vivien, F.D. (2013). Les visions économiques du développement durable : quels enjeux en termes d'évaluation ? In *L'évaluation de la durabilité*, Vivien, F.D., Lepart, J., Marty, P. (eds). Quæ, Versailles.

Vivien, F.D., Nieddu, M., Debref, R. (2019). L'innovation technique : un nouveau paradigme pour le développement soutenable ? In *Développement durable et économie environnementale régionale*, Hamaide, B. and Bruner, S. (eds). Presses de l'Université Saint-Louis, Brussels.

van Weenen, J.C. (1995). Towards sustainable product development. *Journal of Cleaner Production*, 3(1/2), 95–100.

Weyrauch, T. and Herstatt, C. (2017). What is frugal innovation? Three defining criteria. *Journal of Frugal Innovation*, 2(1), 1–17.

Wood, S., Tanner, G., Richardson, B.J. (2010). What ever happened to Canadian Environmental Law? *Ecology Law Quarterly*, 37, 981–1040.

Zaidi-Chtourou, S. and Vernier, M.-F. (2017). Le management de la légitimité : analyse du discours des enseignes spécialisées dans la distribution de produits biologiques. *Décisions Marketing*, 86(2), 67–88.

Index

P, R

S, T

U, W

Other titles from

iSTE

in

Innovation, Entrepreneurship and Management

2023

DARTIGUEPEYROU Carine, SALOFF-COSTE Michel
Futures: The Great Turn
(Innovation and Technology Set – Volume 18)

PEYROUX Élisabeth, RAIMOND Christine, VIEL Vincent, VALIE Émilie
Development and Territorial Restructuring in an Era of Global Change: Theories, Approaches and Future Research Perspectives

SAULAIS Pierre
Knowledge and Ideation: Inventive Knowledge Analysis for Ideation Stimulation
(Innovation and Technology Set – Volume 17)

2022

AOUINAÏT Camille
Open Innovation Strategies
(Smart Innovation Set – Volume 39)

BOUCHÉ Geneviève
Productive Economy, Contributory Economy: Governance Tools for the Third Millennium
(Innovation and Technology Set – Volume 15)

BRUYÈRE Christelle
Caring Management in Health Organizations: A Lever for Crisis Management
(Health and Innovation Set – Volume 3)

HELLER David
Valuation of the Liability Structure by Real Options
(Modern Finance, Management Innovation and Economic Growth Set – Volume 5)

MATHIEU Valérie
A Customer-oriented Manager for B2B Services: Principles and Implementation

MORALES Lucía, DZEVER Sam, TAYLOR Robert
Asia-Europe Industrial Connectivity in Times of Crisis
(Innovation and Technology Set – Volume 16)

NOËL Florent, SCHMIDT Géraldine
Employability and Industrial Mutations: Between Individual Trajectories and Organizational Strategic Planning
(Technological Changes and Human Resources Set – Volume 4)

DE SAINT JULIEN Odile
The Innovation Ecosystem as a Source of Value Creation: A Value Creation Lever for Open Innovation
(Diverse and Global Perspectives on Value Creation Set – Volume 4)

SALOFF-COSTE Michel
Innovation Ecosystems: The Future of Civilizations and the Civilization of the Future
(Innovation and Technology Set – Volume 14)

VAYRE Emilie
Digitalization of Work: New Spaces and New Working Times
(Technological Changes and Human Resources Set – Volume 5)

ZAFEIRIS Konstantinos N, SKIADIS Christos H, DIMOTIKALIS Yannis, KARAGRIGORIOU Alex, KARAGRIGORIOU-VONTA Christina
Data Analysis and Related Applications 1: Computational, Algorithmic and Applied Economic Data Analysis
(Big Data, Artificial Intelligence and Data Analysis Set – Volume 9)
Data Analysis and Related Applications 2: Multivariate, Health and Demographic Data Analysis
(Big Data, Artificial Intelligence and Data Analysis Set – Volume 10)

2021

ARCADE Jacques
Strategic Engineering
(Innovation and Technology Set – Volume 11)

BÉRANGER Jérôme, RIZOULIÈRES Roland
The Digital Revolution in Health
(Health and Innovation Set – Volume 2)

BOBILLIER CHAUMON Marc-Eric
Digital Transformations in the Challenge of Activity and Work: Understanding and Supporting Technological Changes
(Technological Changes and Human Resources Set – Volume 3)

BUCLET Nicolas
Territorial Ecology and Socio-ecological Transition
(Smart Innovation Set – Volume 34)

DIMOTIKALIS Yannis, KARAGRIGORIOU Alex, PARPOULA Christina, SKIADIS Christos H
Applied Modeling Techniques and Data Analysis 1: Computational Data Analysis Methods and Tools
(Big Data, Artificial Intelligence and Data Analysis Set - Volume 7)
Applied Modeling Techniques and Data Analysis 2: Financial, Demographic, Stochastic and Statistical Models and Methods
(Big Data, Artificial Intelligence and Data Analysis Set – Volume 8)

DISPAS Christophe, KAYANAKIS Georges, SERVEL Nicolas, STRIUKOVA Ludmila
Innovation and Financial Markets
(Innovation between Risk and Reward Set – Volume 7)

ENJOLRAS Manon
Innovation and Export: The Joint Challenge of the Small Company
(Smart Innovation Set – Volume 37)

FLEURY Sylvain, RICHIR Simon
Immersive Technologies to Accelerate Innovation: How Virtual and Augmented Reality Enables the Co-Creation of Concepts
(Smart Innovation Set – Volume 38)

GIORGINI Pierre
The Contributory Revolution
(Innovation and Technology Set – Volume 13)

GOGLIN Christian
Emotions and Values in Equity Crowdfunding Investment Choices 2: Modeling and Empirical Study

GRENIER Corinne, OIRY Ewan
Altering Frontiers: Organizational Innovations in Healthcare
(Health and Innovation Set – Volume 1)

GUERRIER Claudine
Security and Its Challenges in the 21st Century
(Innovation and Technology Set – Volume 12)

HELLER David
Performance of Valuation Methods in Financial Transactions
(Modern Finance, Management Innovation and Economic Growth Set – Volume 4)

LEHMANN Paul-Jacques
Liberalism and Capitalism Today

SOULÉ Bastien, HALLÉ Julie, VIGNAL Bénédicte, BOUTROY Éric, NIER Olivier
Innovation in Sport: Innovation Trajectories and Process Optimization
(Smart Innovation Set – Volume 35)

UZUNIDIS Dimitri, KASMI Fedoua, ADATTO Laurent
Innovation Economics, Engineering and Management Handbook 1: Main Themes
Innovation Economics, Engineering and Management Handbook 2: Special Themes

VALLIER Estelle
Innovation in Clusters: Science–Industry Relationships in the Face of Forced Advancement
(Smart Innovation Set – Volume 36)

2020

ACH Yves-Alain, RMADI-SAÏD Sandra
Financial Information and Brand Value: Reflections, Challenges and Limitations

ANDREOSSO-O'CALLAGHAN Bernadette, DZEVER Sam, JAUSSAUD Jacques, TAYLOR Robert
Sustainable Development and Energy Transition in Europe and Asia
(Innovation and Technology Set – Volume 9)

BEN SLIMANE Sonia, M'HENNI Hatem
Entrepreneurship and Development: Realities and Future Prospects
(Smart Innovation Set – Volume 30)

CHOUTEAU Marianne, FOREST Joëlle, NGUYEN Céline
Innovation for Society: The P.S.I. Approach
(Smart Innovation Set – Volume 28)

CORON Clotilde
Quantifying Human Resources: Uses and Analysis
(Technological Changes and Human Resources Set – Volume 2)

CORON Clotilde, GILBERT Patrick
Technological Change
(Technological Changes and Human Resources Set – Volume 1)

CERDIN Jean-Luc, PERETTI Jean-Marie
The Success of Apprenticeships: Views of Stakeholders on Training and Learning (Human Resources Management Set – Volume 3)

DELCHET-COCHET Karen
Circular Economy: From Waste Reduction to Value Creation
(Economic Growth Set – Volume 2)

DIDAY Edwin, GUAN Rong, SAPORTA Gilbert, WANG Huiwen
Advances in Data Science
(Big Data, Artificial Intelligence and Data Analysis Set – Volume 4)

DOS SANTOS PAULINO Victor
Innovation Trends in the Space Industry
(Smart Innovation Set – Volume 25)

GASMI Nacer
Corporate Innovation Strategies: Corporate Social Responsibility and Shared Value Creation
(Smart Innovation Set – Volume 33)

GOGLIN Christian
Emotions and Values in Equity Crowdfunding Investment Choices 1: Transdisciplinary Theoretical Approach

GUILHON Bernard
Venture Capital and the Financing of Innovation
(Innovation Between Risk and Reward Set – Volume 6)

LATOUCHE Pascal
Open Innovation: Human Set-up
(Innovation and Technology Set – Volume 10)

LIMA Marcos
Entrepreneurship and Innovation Education: Frameworks and Tools
(Smart Innovation Set – Volume 32)

MACHADO Carolina, DAVIM J. Paulo
Sustainable Management for Managers and Engineers

MAKRIDES Andreas, KARAGRIGORIOU Alex, SKIADAS Christos H.
Data Analysis and Applications 3: Computational, Classification, Financial, Statistical and Stochastic Methods
(Big Data, Artificial Intelligence and Data Analysis Set – Volume 5)
Data Analysis and Applications 4: Financial Data Analysis and Methods
(Big Data, Artificial Intelligence and Data Analysis Set – Volume 6)

MASSOTTE Pierre, CORSI Patrick
Complex Decision-Making in Economy and Finance

MEUNIER François-Xavier
Dual Innovation Systems: Concepts, Tools and Methods
(Smart Innovation Set – Volume 31)

MICHAUD Thomas
Science Fiction and Innovation Design
(Innovation in Engineering and Technology Set – Volume 6)

MONINO Jean-Louis
Data Control: Major Challenge for the Digital Society
(Smart Innovation Set – Volume 29)

MORLAT Clément
Sustainable Productive System: Eco-development versus Sustainable Development (Smart Innovation Set – Volume 26)

SAULAIS Pierre, ERMINE Jean-Louis
Knowledge Management in Innovative Companies 2: Understanding and Deploying a KM Plan within a Learning Organization
(Smart Innovation Set – Volume 27)

2019

AMENDOLA Mario, GAFFARD Jean-Luc
Disorder and Public Concern Around Globalization

BARBAROUX Pierre
Disruptive Technology and Defence Innovation Ecosystems
(Innovation in Engineering and Technology Set – Volume 5)

DOU Henri, JUILLET Alain, CLERC Philippe
Strategic Intelligence for the Future 1: A New Strategic and Operational Approach
Strategic Intelligence for the Future 2: A New Information Function Approach

FRIKHA Azza
Measurement in Marketing: Operationalization of Latent Constructs

FRIMOUSSE Soufyane
Innovation and Agility in the Digital Age
(Human Resources Management Set – Volume 2)

GAY Claudine, SZOSTAK Bérangère L.
Innovation and Creativity in SMEs: Challenges, Evolutions and Prospects
(Smart Innovation Set – Volume 21)

GORIA Stéphane, HUMBERT Pierre, ROUSSEL Benoît
Information, Knowledge and Agile Creativity
(Smart Innovation Set – Volume 22)

HELLER David
Investment Decision-making Using Optional Models
(Economic Growth Set – Volume 2)

HELLER David, DE CHADIRAC Sylvain, HALAOUI Lana, JOUVET Camille
The Emergence of Start-ups
(Economic Growth Set – Volume 1)

HÉRAUD Jean-Alain, KERR Fiona, BURGER-HELMCHEN Thierry
Creative Management of Complex Systems
(Smart Innovation Set – Volume 19)

LATOUCHE Pascal
Open Innovation: Corporate Incubator
(Innovation and Technology Set – Volume 7)

LEHMANN Paul-Jacques
The Future of the Euro Currency

LEIGNEL Jean-Louis, MÉNAGER Emmanuel, YABLONSKY Serge
Sustainable Enterprise Performance: A Comprehensive Evaluation Method

LIÈVRE Pascal, AUBRY Monique, GAREL Gilles
Management of Extreme Situations: From Polar Expeditions to Exploration-Oriented Organizations

MILLOT Michel
Embarrassment of Product Choices 2: Towards a Society of Well-being

N'GOALA Gilles, PEZ-PÉRARD Virginie, PRIM-ALLAZ Isabelle
Augmented Customer Strategy: CRM in the Digital Age

NIKOLOVA Blagovesta
The RRI Challenge: Responsibilization in a State of Tension with Market Regulation
(Innovation and Responsibility Set – Volume 3)

PELLEGRIN-BOUCHER Estelle, ROY Pierre
Innovation in the Cultural and Creative Industries
(Innovation and Technology Set – Volume 8)

PRIOLON Joël
Financial Markets for Commodities

QUINIOU Matthieu
Blockchain: The Advent of Disintermediation

RAVIX Joël-Thomas, DESCHAMPS Marc
Innovation and Industrial Policies
(Innovation between Risk and Reward Set – Volume 5)

ROGER Alain, VINOT Didier
Skills Management: New Applications, New Questions
(Human Resources Management Set – Volume 1)

SAULAIS Pierre, ERMINE Jean-Louis
Knowledge Management in Innovative Companies 1: Understanding and Deploying a KM Plan within a Learning Organization
(Smart Innovation Set – Volume 23)

SERVAJEAN-HILST Romaric
Co-innovation Dynamics: The Management of Client-Supplier Interactions for Open Innovation
(Smart Innovation Set – Volume 20)

SKIADAS Christos H., BOZEMAN James R.
Data Analysis and Applications 1: Clustering and Regression, Modeling-estimating, Forecasting and Data Mining
(Big Data, Artificial Intelligence and Data Analysis Set – Volume 2)
Data Analysis and Applications 2: Utilization of Results in Europe and Other Topics
(Big Data, Artificial Intelligence and Data Analysis Set – Volume 3)

UZUNIDIS Dimitri
Systemic Innovation: Entrepreneurial Strategies and Market Dynamics

VIGEZZI Michel
World Industrialization: Shared Inventions, Competitive Innovations and Social Dynamics
(Smart Innovation Set – Volume 24)

2018

BURKHARDT Kirsten
Private Equity Firms: Their Role in the Formation of Strategic Alliances

CALLENS Stéphane
Creative Globalization
(Smart Innovation Set – Volume 16)

CASADELLA Vanessa
Innovation Systems in Emerging Economies: MINT – Mexico, Indonesia, Nigeria, Turkey
(Smart Innovation Set – Volume 18)

CHOUTEAU Marianne, FOREST Joëlle, NGUYEN Céline
Science, Technology and Innovation Culture
(Innovation in Engineering and Technology Set – Volume 3)

CORLOSQUET-HABART Marine, JANSSEN Jacques
Big Data for Insurance Companies
(Big Data, Artificial Intelligence and Data Analysis Set – Volume 1)

CROS Françoise
Innovation and Society
(Smart Innovation Set – Volume 15)

DEBREF Romain
Environmental Innovation and Ecodesign: Certainties and Controversies
(Smart Innovation Set – Volume 17)

DOMINGUEZ Noémie
SME Internationalization Strategies: Innovation to Conquer New Markets

ERMINE Jean-Louis
Knowledge Management: The Creative Loop
(Innovation and Technology Set – Volume 5)

GILBERT Patrick, BOBADILLA Natalia, GASTALDI Lise,
LE BOULAIRE Martine, LELEBINA Olga
Innovation, Research and Development Management

IBRAHIMI Mohammed
Mergers & Acquisitions: Theory, Strategy, Finance

LEMAÎTRE Denis
Training Engineers for Innovation

LÉVY Aldo, BEN BOUHENI Faten, AMMI Chantal
Financial Management: USGAAP and IFRS Standards
(Innovation and Technology Set – Volume 6)

MILLOT Michel
Embarrassment of Product Choices 1: How to Consume Differently

PANSERA Mario, OWEN Richard
Innovation and Development: The Politics at the Bottom of the Pyramid
(Innovation and Responsibility Set – Volume 2)

RICHEZ Yves
Corporate Talent Detection and Development

SACHETTI Philippe, ZUPPINGER Thibaud
New Technologies and Branding
(Innovation and Technology Set – Volume 4)

SAMIER Henri
Intuition, Creativity, Innovation

TEMPLE Ludovic, COMPAORÉ SAWADOGO Eveline M.F.W.
Innovation Processes in Agro-Ecological Transitions in Developing Countries
(Innovation in Engineering and Technology Set – Volume 2)

UZUNIDIS Dimitri
Collective Innovation Processes: Principles and Practices
(Innovation in Engineering and Technology Set – Volume 4)

VAN HOOREBEKE Delphine
The Management of Living Beings or Emo-management

2017

AÏT-EL-HADJ Smaïl
The Ongoing Technological System
(Smart Innovation Set – Volume 11)

BAUDRY Marc, DUMONT Béatrice
Patents: Prompting or Restricting Innovation?
(Smart Innovation Set – Volume 12)

BÉRARD Céline, TEYSSIER Christine
Risk Management: Lever for SME Development and Stakeholder Value Creation

CHALENÇON Ludivine
Location Strategies and Value Creation of International Mergers and Acquisitions

CHAUVEL Danièle, BORZILLO Stefano
The Innovative Company: An Ill-defined Object
(Innovation between Risk and Reward Set – Volume 1)

CORSI Patrick
Going Past Limits To Growth

D'ANDRIA Aude, GABARRET Inés
Building 21st Century Entrepreneurship
(Innovation and Technology Set – Volume 2)

DAIDJ Nabyla
Cooperation, Coopetition and Innovation
(Innovation and Technology Set – Volume 3)

FERNEZ-WALCH Sandrine
The Multiple Facets of Innovation Project Management
(Innovation between Risk and Reward Set – Volume 4)

FOREST Joëlle
Creative Rationality and Innovation
(Smart Innovation Set – Volume 14)

GUILHON Bernard
Innovation and Production Ecosystems
(Innovation between Risk and Reward Set – Volume 2)

HAMMOUDI Abdelhakim, DAIDJ Nabyla
Game Theory Approach to Managerial Strategies and Value Creation
(Diverse and Global Perspectives on Value Creation Set – Volume 3)

LALLEMENT Rémi
Intellectual Property and Innovation Protection: New Practices and New Policy Issues
(Innovation between Risk and Reward Set – Volume 3)

LAPERCHE Blandine
Enterprise Knowledge Capital
(Smart Innovation Set – Volume 13)

LEBERT Didier, EL YOUNSI Hafida
International Specialization Dynamics
(Smart Innovation Set – Volume 9)

MAESSCHALCK Marc
Reflexive Governance for Research and Innovative Knowledge
(Responsible Research and Innovation Set – Volume 6)

MASSOTTE Pierre
Ethics in Social Networking and Business 1: Theory, Practice and Current Recommendations
Ethics in Social Networking and Business 2: The Future and Changing Paradigms

MASSOTTE Pierre, CORSI Patrick
Smart Decisions in Complex Systems

MEDINA Mercedes, HERRERO Mónica, URGELLÉS Alicia
Current and Emerging Issues in the Audiovisual Industry
(Diverse and Global Perspectives on Value Creation Set – Volume 1)

MICHAUD Thomas
Innovation, Between Science and Science Fiction
(Smart Innovation Set – Volume 10)

PELLÉ Sophie
Business, Innovation and Responsibility
(Responsible Research and Innovation Set – Volume 7)

SAVIGNAC Emmanuelle
The Gamification of Work: The Use of Games in the Workplace

SUGAHARA Satoshi, DAIDJ Nabyla, USHIO Sumitaka
Value Creation in Management Accounting and Strategic Management: An Integrated Approach
(Diverse and Global Perspectives on Value Creation Set –Volume 2)

UZUNIDIS Dimitri, SAULAIS Pierre
Innovation Engines: Entrepreneurs and Enterprises in a Turbulent World
(Innovation in Engineering and Technology Set – Volume 1)

2016

BARBAROUX Pierre, ATTOUR Amel, SCHENK Eric
Knowledge Management and Innovation
(Smart Innovation Set – Volume 6)

BEN BOUHENI Faten, AMMI Chantal, LEVY Aldo
Banking Governance, Performance And Risk-Taking: Conventional Banks Vs Islamic Banks

BOUTILLIER Sophie, CARRÉ Denis, LEVRATTO Nadine
Entrepreneurial Ecosystems
(Smart Innovation Set – Volume 2)

BOUTILLIER Sophie, UZUNIDIS Dimitri
The Entrepreneur
(Smart Innovation Set – Volume 8)

BOUVARD Patricia, SUZANNE Hervé
Collective Intelligence Development in Business

GALLAUD Delphine, LAPERCHE Blandine
Circular Economy, Industrial Ecology and Short Supply Chains
(Smart Innovation Set – Volume 4)

GUERRIER Claudine
Security and Privacy in the Digital Era
(Innovation and Technology Set – Volume 1)

MEGHOUAR Hicham
Corporate Takeover Targets

MONINO Jean-Louis, SEDKAOUI Soraya
Big Data, Open Data and Data Development
(Smart Innovation Set – Volume 3)

MOREL Laure, LE ROUX Serge
Fab Labs: Innovative User
(Smart Innovation Set – Volume 5)

PICARD Fabienne, TANGUY Corinne
Innovations and Techno-ecological Transition
(Smart Innovation Set – Volume 7)

2015

CASADELLA Vanessa, LIU Zeting, DIMITRI Uzunidis
Innovation Capabilities and Economic Development in Open Economies
(Smart Innovation Set – Volume 1)

CORSI Patrick, MORIN Dominique
Sequencing Apple's DNA

CORSI Patrick, NEAU Erwan
Innovation Capability Maturity Model

FAIVRE-TAVIGNOT Bénédicte
Social Business and Base of the Pyramid

GODÉ Cécile
Team Coordination in Extreme Environments

MAILLARD Pierre
Competitive Quality and Innovation

MASSOTTE Pierre, CORSI Patrick
Operationalizing Sustainability

MASSOTTE Pierre, CORSI Patrick
Sustainability Calling

2014

DUBÉ Jean, LEGROS Diègo
Spatial Econometrics Using Microdata

LESCA Humbert, LESCA Nicolas
Strategic Decisions and Weak Signals

2013

HABART-CORLOSQUET Marine, JANSSEN Jacques, MANCA Raimondo
VaR Methodology for Non-Gaussian Finance

2012

DAL PONT Jean-Pierre
Process Engineering and Industrial Management

MAILLARD Pierre
Competitive Quality Strategies

POMEROL Jean-Charles
Decision-Making and Action

SZYLAR Christian
UCITS Handbook

2011

LESCA Nicolas
Environmental Scanning and Sustainable Development

LESCA Nicolas, LESCA Humbert
Weak Signals for Strategic Intelligence: Anticipation Tool for Managers

MERCIER-LAURENT Eunika
Innovation Ecosystems

2010

SZYLAR Christian
Risk Management under UCITS III/IV

2009

COHEN Corine
Business Intelligence

ZANINETTI Jean-Marc
Sustainable Development in the USA

2008

CORSI Patrick, DULIEU Mike
The Marketing of Technology Intensive Products and Services

DZEVER Sam, JAUSSAUD Jacques, ANDREOSSO Bernadette
Evolving Corporate Structures and Cultures in Asia: Impact of Globalization

2007

AMMI Chantal
Global Consumer Behavior

2006

BOUGHZALA Imed, ERMINE Jean-Louis
Trends in Enterprise Knowledge Management

CORSI Patrick *et al.*
Innovation Engineering: the Power of Intangible Networks

Printed and bound by CPI Group (UK) Ltd, Croydon, CR0 4YY
13/09/2026
14961041-0002